House Beautiful

GLASS HOUSE

House Beautiful's
GLASS HOUSE
The Art of Decorating with Light

C.J. PETERSEN

HEARST BOOKS
A Division of Sterling Publishing Co., Inc.
NEW YORK

Library of Congress Cataloging-in-Publication Data

Petersen, CJ

House beautiful's Glass house: the art of decorating with light / C.J. Petersen
    p. cm.
Includes index.
ISBN-13: 978-1-58816-505-3
ISBN-10: 1-58816-505-1
1. Glass in interior decoration. 2. Light in architecture.
3. Glass construction. II. Title:

NK2115.5.L5P48 2006
728--dc22   2006041082

10 9 8 7 6 5 4 3 2 1

Published by Hearst Books
A Division of Sterling Publishing Co., Inc.
387 Park Avenue South, New York, NY 10016

House Beautiful and Hearst Books are trademarks of Hearst Communications, Inc.

www.housebeautiful.com

For information about custom editions, special sales, premium and corporate purchases, please contact Sterling Special Sales Department at 800-805-5489 or specialsales@sterlingpub.com.

Distributed in Canada by Sterling Publishing
c/o Canadian Manda Group, 165 Dufferin Street
Toronto, Ontario, Canada M6K 3H6

Distributed in Australia by Capricorn Link (Australia) Pty. Ltd.
P.O. Box 704, Windsor, NSW 2756 Australia

Manufactured in China

Sterling ISBN 13: 978-1-58816-505-3
       ISBN 10: 1-58816-505-1

FOREWORD   8

INTRODUCTION   12

1. WINDOWS   16

2. DOORS & ENTRYWAYS   48

3. INTERIOR DOORS & WINDOWS   76

4. SKYLIGHTS   100

5. GLASS WALLS & DIVIDERS   122

6. GLASS ACCENTS   150

GLOSSARY   174

PHOTOGRAPHY CREDITS   180

INDEX   184

Contents

FOREWORD

Looking through the many stunning photos we review

for each issue of House Beautiful, we're invariably struck by just how essential natural light is to interior design. The homes and rooms we show in the magazine would be radically diminished without that light. It's not merely that a wash of sunlight activates the color of wall paint or brings out the details of furniture and accessories; it actually creates a mood in a room. We've all felt the exhilaration of waking up to a sun-filled home on a bright spring morning. And there's something about the filtered light of an overcast day that makes you want to start a fire and gravitate toward the overstuffed, well-worn comfortable corners of your home. Natural light enriches our living spaces in ways artificial light could never hope to. How much natural light your interior receives and where it's directed is a matter of the type of glass openings you incorporate into your home.

We think most people underestimate the role glass plays in a home because the material itself is usually invisible. And the ways in which glass affects the home environment are often very subtle, if compelling and profound. In one editor's house, the young daughters gravitate toward one cozy corner of the living room because a pair of windows on intersecting walls flood the corner with sun throughout the day. It's a great place to read or just spend a while daydreaming. With the curtains drawn back in the morning, the sunlight drapes itself across the sofa like a honey-colored afghan. The scene just spells comfort; it's no wonder the girls are

drawn to it. The same is true of the bay window in another's kitchen. In the space once filled by a pair of modest casement windows, there's now a panoramic view of a garden. Although it's wonderful to start each morning with such a nice vista, they rarely think about the configuration of glass that makes it possible.

But glass can be more than an avenue for light or the lens for a view; it can be an architectural focal point in its own right. One editor recently visited a friend's home for a cocktail party. Walking up to their front door he was amazed at how the light from the entryway chandelier literally sparkled through the beveled door insert and sidelites. The facets of the beveled glass created an enchanting prismatic effect, with quick-changing, multicolored flashes greeting them as they approached. You couldn't ask for a more powerful and elegant first impression of a home. That decorative power doesn't stop at the front door either. An interior wall of small panes arranged in an asymmetrical design creates an attention-getting visual to rival art on the walls. A glass countertop surrounding a kitchen island creates a showcase centerpiece in the space.

Unfortunately, in many homes, these effects are accidental if they happen at all. Many homeowners merely "see" glass as a necessary architectural fixture. Although they will gladly change the paint on the walls, add architectural details such as wainscoting and special effects such as imported tile, they don't take steps to control the light that controls the appearance of all those elements. We hope this book changes that; we hope it opens the reader's eyes literally and figuratively. In the following pages, you'll discover the amazing functional and aesthetic potential inherent in glass, illustrated in hundreds of pictures and discussed in straightforward chapters organized by types of glass installations. You'll find special sections that break down different types of glass structures with the pros and cons of each, and we've included style guides throughout to help you choose exactly the right type of glass for your home, décor and tastes. We think you'll be especially interested in the section on art glass at the end of each chapter. Each one of these sections deals with a different glassworking technique, from stained to beveled glass and beyond. The art is original and unique, designed for specfic spaces; choosing glass art can be a very special way to create a signature look in your home. But no matter what type of glass you choose, or how you use it in your home, we think this book will help you see it in a whole new light.

The Editors of House Beautiful

11

FOREWORD

INTRODUCTION

Light is the most important interior design element

in any home. If you doubt this, walk into a room and close your eyes. That's how your interior design looks without light. Light brings every other design element into perspective, from architectural details and decorations to all your furnishings. And although you could light your home with an artificial light source, nothing can replace the rich and complex character, and exceptional exposure, of natural light.

Natural light is so important that we could not live without it. It is essential if our bodies are to produce crucial vitamin D. In fact, as much as twenty percent of the U.S. population suffers from a lack of natural light, manifested in the condition known as seasonal affective disorder (SAD). Characterized by depression, lethargy, fatigue and other symptoms, people with SAD even turn to special lamps with bulbs that replicate the wavelengths of the sun. Studies have shown that students do better when their classes are held in natural light. There's simply no way to do without natural light. Such is the power of the sun.

Thanks to glass, homeowners can exploit this plentiful and crucial resource in any number of ways. Glass openings are the means by which we bring natural light into our living spaces, and the type of glass openings we include in our homes determines how much light enters, where it goes, and how it affects the interior design. This isn't just a matter of how big a window is, either. Where a window and glass door is positioned is just as important, and can radically affect the amount of direct light different areas receive. That's why *Glass House* begins with a discussion of lighting exposures—once you've learned the

differences between light from different directions, you'll be much better equipped to make the most of that light.

But as important as natural light is, the benefits of architectural glass only begin with how it controls the light. Glass openings increase the visual space of rooms, making interior spaces seem larger and integrating those spaces with the surrounding environment. Exterior windows and doors offer views to the outside world and a way to bring in fresh air. These openings provide a changing tableau we can enjoy throughout the seasons, and answers to more pragmatic concerns, such as gauging the weather and determining what it is that's making the dog bark. This book presents in-depth advice on where and how to position windows and doors to best serve all these visual roles.

Beyond the function of glass in helping us see, lies its decorative potential. Glass is a unique material that can be altered in surprising and numerous ways. This can be as simple as innovative framing that arranges basic glass panes in eye-catching asymmetrical patterns, or as complex as beveled glass leaded into an ornate design that fills a door with a vibrant graphic combination of line, texture, and light. This book explores the many possibilities, offering comprehensive information on integrating the many types of glass treatments into the home. Each chapter describes ways to get the most out of any glass application without letting it overpower other design elements.

The chapters themselves are organized around the way glass is used rather than by room, because the fundamental principles of how glass affects a space often hold true regardless of the space. But you'll also find information about how lighting and glass installations affect the function of the room. You should select and focus on the installation that makes the most sense for your home and your preferences. For instance, you may like your exterior windows and doors just the way they are, but want to bring more light into the largest rooms of the house. In which case, a skylight might be the change your interior is crying out for. Or perhaps you're not convinced that it's the time to commit to a significant architectural change, and would prefer to wander through the more modest options in the chapter on glass accents. You'll find all those alternatives explored in depth.

You'll also find options you may never have heard about before, such as glass floor tiles, or unusual art glass variations that can help create a unique look in your home while modulating the light in stunning and astonishing ways. But plain or grand, art or simple glazed aperture, in the end you should choose the glass that fits not just an opening, but your home, your natural light requirements, and your tastes.

Windows

▷ Floor-to-ceiling windows open
this space out, visually increasing
the space to beyond the walls.
The large windows also increase
light exposure.

Windows are the key source of natural light in a home.

Their sizes and positioning affect the quality of light in a given room throughout the day, and how other design elements—from the paint on the walls to
the furniture—will appear at any point in time. By conducting light into the
home, windows define living spaces. This is a lofty enough role to play, but
they do even more.

Windows connect us to the world beyond our walls. We look out and
engage as we let the world look in. Lighted windows are one way we tell every-
one we are at home, revealing the warmth and character of both house and
homeowner. They are visual passageways that connect our interior spaces with
nature and the community around us.

They also serve as design elements themselves. Framed glass breaks up long
walls, providing interesting visual relief. An unusual shape or graphic combi-

nation of window styles makes for a dynamic visual presence. The framing pieces—outer trim, sash, and mullions—are themselves design elements, adding line, color and texture to an interior design scheme.

No less important are the structural roles windows play. They supply the ventilation for a home, and today's insulated windows often provide as much insulation value as walls do. And windows are the best way for occupants to visually scan the area around a house, making them invaluable for home security. But the first consideration in choosing the type of windows and their placement is light exposure. Any window is ultimately servant to the light.

▼ Side-by-side windows add to the sense of balance in the living room composition, with simple window treatments that allow a maximum of light to enter while complementing the interior decor.

Window Types

The key difference that defines window styles is how they open. This determines what architectural style the window is best suited for, how convenient it will be to operate, and how easy it is to clean.

Fixed Glass: Often called "picture windows," these encompass any non-opening window. You'll find them in a variety of sizes and shapes. Large picture windows are generally used to frame a dramatic view or where a large opening is natural to the architecture. Smaller fixed-glass windows are good choices as accents to complement other types of windows. The vast array of fixed-glass window styles means there is one for just about any architectural style. They are also excellent choices for inaccessible areas, where a moving sash would be difficult if not impossible to open.

Single Hung/Double Hung: This ubiquitous design is divided into upper and lower sashes. The lower sash opens by sliding up in front of the upper sash. In single-hung windows, only the lower sash moves; in double hung models, both sashes can be slid up or down. This type of window is suitable to a wide range of home styles, from colonial, Victorian, and other traditional forms, to modern lofts where oversized units serve as floor-to-ceiling windows. Different divider and trim styles can be used to match the window to the home's design. Double-hung units are also a great choice where a swing-out window would pose a problem, such as over a walkway or patio.

Casement: Excellent at catching breezes and directing airflow into the home, casement windows are hinged along one side, with a crank that opens the window out. Casement windows are a contemporary look that complements simple, linear architectural styles such as Prairie School, Tudor, ranch and modern homes. This type of window is ideal in areas where access to the window is blocked—such as over a kitchen sink—and it would be difficult to open a double-hung or gliding window.

Gliding: Also called "sliding," these windows feature side-by-side panels that open and close by sliding horizontally in front of one another. Like double-hung windows, they work great in spaces where a swing-out window would be in the way. Their simple look and horizontal orientation make these windows good choices for more contemporary homes, or buildings with minimal architectural ornamentation.

Awning: As the name implies, awning windows open out from the bottom. They are regularly combined with other windows to create attractive groupings and provide ventilation in combination with fixed windows. They are used as transom windows in traditional and period structures such as Craftsman Style, Tudor, or Victorian, and in contemporary or modern homes grouped with other types.

Bay: This is not one window, but a collection of three or more windows arranged in a projection that juts out from the interior space. Bay windows are not actually curved—their shapes are created by the individual windows meeting edge-to-edge at angles. The construction captures more light due to multiple angles of exposure. These carry significant visual weight and should be carefully balanced by other architectural elements so that they don't overwhelm the exterior appearance of the home.

Bow: Similar to a bay window, a bow window is four or more windows arranged in a contiguous arc. The construction creates a smooth curve that can range from a simple arc to a three-quarter circle. The look is extremely adaptable, and bow windows can be used with just about any architectural style if carefully planned into the layout.

FABULOUS FENESTRATION

Fenestration is the architectural term used to describe the arrangement and proportion of windows in a structure. Although often used to describe the overall composition of all openings in a wall or building, the term actually refers only to windows, and is derived from the Latin word "fenestra," which means window.

Just the same, the placement of windows bears a relationship to other openings such as doors. For instance, if you decide to add a bay or bow window in a living room alongside the entryway, the relationship between the two openings will be extremely important. A large, floor-to-ceiling bay or bow window may make a dramatic interior statement in the living room, opening the room up and greatly increasing natural light. But viewed from the exterior, that window may overpower the architecture. It may actually be larger than the entryway and undermine the overall effect of the home's façade.

More commonly, you'll have to judge the general composition of a group of windows along a wall, or across a home's façade. Simple symmetry is often the easiest and safest way to create a pleasing graphic pattern. An example of this is a row of identical windows positioned at the same height and spaced regularly along a wall. Asymmetrical window placement can add excitement and visual interest to a home's interior and exterior views, but this fenestration style can also go horribly wrong. Whatever the composition, effective fenestration requires that you consider both interior and exterior views.

Myriad other concerns will also affect decisions concerning fenestration. Providing adequate ventilation to the interior is one. Cross-ventilation can be radically affected by an imbalance of windows on two sides of a structure. In addition, local building code requirements may also control what you can and can't do with windows. Certainly, if your home is a historic structure or if it is located within a designated historical district, local, state or federal regulations may restrict what changes you can make regarding window placement.

That's why it's often best to call on a professional such as an architect to determine appropriate fenestration. Experts have experience in balancing window placement with other architectural elements, and will be more aware of code requirements, historic landmark regulations, and architectural style guidelines. If you are building a new home, you'll naturally be consulting an architectural professional who can give you exciting options for the plan you are considering. But even if you're installing a single window, an architectural consultation is a good idea.

▶ Fenestration can be an opportunity to make a dramatic statement within the architecture of your home. Here, two stunning triangular fixed-glass windows accentuate the extreme angle of the roof and enliven the façade of the structure.

LIGHT RULES

Windows are first and foremost gateways for light. Nothing changes the visual character of a room so much as a flood of sunlight pouring through a well-placed window. But as enchanting as natural light is, it also presents a design challenge because it's constantly changing. Its quality or appearance varies with the seasons and even the time of day. Even the color of sunlight changes; it can be an intense, pure white radiance, or a soft amber glow. It can be subtle or harsh, diffuse or concentrated.

Natural light is also colored by environment. For instance, optical effects caused by the ocean's surface give the sunlight that falls on California's coastline a distinctly rich golden hue. As esoteric as this may sound, the color of light is a key ingredient in interior design. Cool blue light renders details accurately to the eye, making interior colors and shapes crisp and realistic. Rich yellow sunlight tends to wash out beige and neutral tones, and softens lines and details.

The direction a window faces radically affects the type of light a room receives, as do the size and shape of the window. Northern exposures offer cool, blue light, while southern exposures present a fuzzier golden light. Even nearby structures such as trees or other buildings will play a part in the type of light that enters a room. Whatever the color or character, sunlight transmits through a window one of two ways: as indirect "ambient" light, or strong, direct light. The type of light a room receives is determined by the window's exposure to the sun.

Exposing Exposures

Light travels a straight path; a given window only receives direct light when the sun is in line with it. Consequently, designing for light means dealing with the

24

Silent Film Star

Window films are coatings designed to moderate sunlight coming through a window. Films allow natural light to pass through the window, while blocking out harmful UV rays and much of the sun's radiant heat. This prevents the light from fading furniture fabric and other surfaces, stops a room with many windows from becoming a "hothouse," and reduces glare spots in TVs and computer monitors. Window film can also be used on the inside of the window to stop the loss of heat or air-conditioned air from the interior.

GLASS FACTS

The almost directly western exposure of this bedroom bow window ensures that the room is shaded in early morning for sleeping, and bright and sunny through the afternoon.

White surfaces provide a canvas for the play of light and shadow pattern as the sun moves through the sky. This effect can create a dynamic visual presence in a room—knowing your home's lighting exposures gives you the tools to control the effect.

basic principle of a spinning Earth: the sun rises in the east and sets in the west, and it sits higher in the sky for longer in summer. Windows facing due east get strong, direct morning light, but will experience ambient light for the rest of the day. Those facing directly west receive afternoon light. South-facing windows are exposed to a hotter, more yellow light, while northern exposures experience a cooler, weaker blue light. Traditionally, the most desirable exposure for sunlight lovers is southwest.

A window's exposure will impact design decisions. For instance, in an east-facing bedroom, unwanted morning light will be a problem unless the windows are outfitted with heavy drapes, blackout shades, or other significant window treatment. The sunlight through a plate glass window with a southwestern exposure will be direct and strong through most of the day. You'll either need to diffuse the light with gauzy curtains, or use a window film to keep furniture and fabrics from fading. You should also consider the nighttime exposure when positioning windows—strong moonlight can be as disruptive to sleep as other natural light.

Whatever direction windows face, the light striking them is also affected by other structures, such as taller adjacent houses that block light, or a large leafy tree that filters light (at least while the leaves are on the branches). Given the

▷ A gable-end fixed-glass window is placed in balance with the row of three double-hung windows below it. The high placement of the window allows light to penetrate deeper into the kitchen—a boon for potted plants that have no windowsill on which to sit.

◁ A simple row of vertical frame elements dividing a long picture window creates captivating linear shadows that move through the spare space of this modern home throughout the day.

number of factors that influence the quality and intensity of light through your windows, you need to carefully assess the exposure in any room you're decorating—whether you're planning on changing the windows or not.

This awareness will drive design decisions. For instance, use white paint for walls, ceilings, furniture or floors to create a wonderful canvas for interesting shadows and patterns. This is also a good way to increase the apparent light in a dimly lit room. Gloss paint and wood finishes are a good idea in a room that receives little direct light or in which the windows are small. Leave wood floors uncarpeted to take advantage of the soft diffuse glow of sunlight washing over them throughout the day.

Where sunlight is strong and intense, you can moderate it with surfaces that absorb light, such as unfinished stone or matte wall paints. Avoid reflective surfaces such as glass or polished wood in rooms that receive direct light for much of the day. Or take advantage of strong light in a room by painting walls in jewel tones or introducing sun-loving houseplants.

Position interior design features to take advantage of your light exposure. Put a sculpture on a shelf that receives light throughout the day to highlight the art. Place a shade-loving plant in a part of the room that lies in shadow, but use a pot with a high-gloss finish to emphasize whatever light reaches the area. When designing for light and glass in the home, it's important to think about where the light won't be as much as where it will.

WINDOW PLACEMENT

Using windows as dynamic graphic elements in an interior design scheme is an exercise in positioning. Existing windows require that furnishings and decorative accents be positioned to take advantage of the light and where it falls. If your home has been built with sound architectural principles and thoughtful fenestration, the existing window configuration will provide the light and graphic appeal you require. Note whether natural light is well distributed throughout the space, and if the composition of windows along the walls is balanced and pleasing to the eye.

Where the existing window configuration creates uneven and ineffective lighting, or seems in odd proportion to the space and architecture, you should consider adding new windows. Adding a window or windows allows you to change the graphic appeal, lighting characteristics, and how the room is viewed and "read."

◀ West-facing windows serve this bedroom well. The room lies in shade until afternoon, ensuring bright mornings don't bring a rude awakening. The third, higher window creates an interesting triad composition, and allows light to penetrate deep into the room.

29

Clean Like a Pro

Many professional window cleaners use an incredibly basic formula to quickly, efficiently, and easily clean windows. Add half a teaspoon of liquid dish soap to a half-gallon of water. Apply the mix to the window with a sponge. If the window is a large undivided pane, use a squeegee to remove the soap. For smaller divided windows, use a soft cotton cloth or old towel to dry the window. The soap not only cleans normal dust and dirt, it also removes stubborn grime and oils, such as fingerprints. It's also a much cheaper option than commercially available window cleaners.

Of course new construction offers great flexibility for creating an entirely fluid window layout, one that provides maximum graphic appeal and still optimizes available light. However, in most cases the home already exists. This means that the vast majority of homeowners will be working with an existing window configuration.

Existing Windows

Designing around existing windows involves balancing the practical aspects of how the room is used and the need for natural lighting against the aesthetic considerations of how that light affects the interior design. Because the windows themselves cannot be moved, this involves arranging furniture and decorative accents to take advantage of the windows.

The first step is to determine the light exposure throughout the day for whatever room you're redecorating. Although where you place furniture will often be a matter of where things fit and where you logically want to position them, you can take advantage of natural light to highlight features such as an elegant side table, framed photos on a wall, or exotic houseplants in unique lacquered pots.

One of the best ways to design around existing windows is to create a "light stage" with a furniture grouping positioned to optimize the natural light in a given room. This can be a comfortable reading chair and side table placed in the sunlit corner of a bedroom or living room, or a window seat or table in the kitchen, positioned to bask in the morning sun while you drink your coffee and read the paper.

New Windows

Adding new windows can be a uniquely powerful way for homeowners to put their own signature on a home. The change can be as modest as updating the style of double-hung windows by replacing old, drafty units with new units that include stylish divided-lite grills, or as radical as crafting new openings for windows that will change the appearance of the structure inside and out.

The incredible number of window options opens up a treasure chest of design possibilities. Mixing and matching different types of windows can be a way to construct interesting graphic patterns while maximizing available light. The right window in the right place can add scale and dimension to a room, illuminate dark corners or nooks, determine what the view from the room will be, and function as a decorative element in and of itself.

▶ As stunning as a work of art, this oval window sports a spider-web grille and traditional colonial frame. The window trim is painted bright white to stand out against the red wall.

▼ The focal point of an otherwise unremarkable hallway is a simple, bright, double-hung window. The natural light is amplified by walls that are painted gloss white and a wood floor stained in a light satin finish.

30

WINDOWS

SHAPE: The days of choosing between square or rectangular windows are thankfully long gone. Contemporary prefabricated windows are available in hexagons, ovals, circles, triangles, arch-top, clip-corner, quarter- or half-round, elliptical and other shapes. If that mind-boggling array doesn't meet your needs, most manufacturers offer custom services to create a window to the size and shape you desire.

Windows with unusual shapes, such as hexagons or arch tops, work best as focal points. Place them where they will draw the most attention. The shapes of windows also provide ways to reinforce the thematic lines of the architecture. For instance, a clip-corner window on an end wall can echo the lines of a cathedral ceiling. Triangular windows are often used in gable-end openings for this purpose.

Unusual shaped windows will also have a powerful effect on exterior appearance, and how the home will look from the outside is key in choosing window shapes. Be careful not to overwhelm the exterior appearance with a window that lets a lot of light in, but is grossly out of proportion to other exterior architectural features.

PERSPECTIVE: If you've ever been in a large room with small windows, you know that windows have the ability to define visual space. For instance, in a room with high ceilings, your furnishings may look squat in relation to the rise of the walls. Fixed-glass windows positioned high on a wall—or triangular gable-end windows—will bring the interior into scale, diminishing the overwhelming visual height. A long thin room can be visually opened out by using a row of windows along the longer walls. Small rooms will seem bigger with windows on every exterior wall.

Where perspective is concerned, few installations will have as dramatic an influence as corner windows. Where two or more windows meet in a corner of a room, they create an unbroken glass wedge that opens the room outward. This allows the eye to follow the room's lines out into the space beyond the window. The effect works well day or night as long as the windows are not covered with drapes or curtains. Corner windows make a room seem much larger, and the configuration offers double exposures of sunlight—which can translate into a well-lit room throughout the entire day.

COMPOSITIONS: Lines and negative space have always been powerful tools in the interior designer's repertoire. Windows grouped in graphic patterns can create attention-getting linework and a stark, interesting contrast between blank wall space and the view beyond the walls. Generally, window groups

◁ Fixed-glass windows positioned in rows over double-hung windows give the room scale, bringing the high ceiling into proper perspective with the furnishings.

▽ A composition of arched and half-arched windows tops an impressive triad of windows with decorative dividers. The window shape and trim elements yield as much graphic appeal as the view beyond.

Translucent white Roman shades provide absolute privacy without losing the light in this small bathroom.

include both fixed-glass and opening windows, to create a compelling mix of sizes and shapes while maintaining ventilation. Window compositions can be complex, such as a group of five or six windows creating a collage in a living room. Or they can be simple, as with a row of small windows positioned directly over a row of double-hung windows. The most common window combinations are symmetrical, providing a pleasing balance to the eye. However, great visual interest and graphic tension can be achieved with an asymmetrical design. For example, three windows on one end of a wall might be matched by two windows in the same amount of space on the other end of the wall.

Window Dressing

Window treatments can affect both the light transmission and look of windows. Your choice of window treatment—including whether to use them or not—should be based on three factors: desired sunlight transmission, the view in and out, and design style.

Desired transmission is simply the amount of light you want coming in through the window. If the light is ideal at all times through the day and night, you don't need a window treatment unless you would like to use it as a design element. If you want to moderate the light at times during the day, you'll probably want to choose filmy curtains with a level of transparency or adjustable blinds or shutters. If blocking the light entirely is a goal, select from shades, blinds, or opaque, lined drapes.

There are two types of view to consider with any window and window treatment—the view in and the view out. If your window looks out on a gorgeous scene, any window treatment should be minimal, so as not to impede any part of the view. For instance, a plate glass window overlooking a private, beautifully designed, four-season garden probably creates all the visual impact you could want without the help of a window treatment. The view into the home is basically a matter of privacy. In rooms such as living rooms, where privacy is most likely not an issue (unless there are overriding security concerns or personal preferences), the choice of window treatment is more likely to be guided by light transmission. But in a bathroom or bedroom, the window treatments serve an important privacy role, and should completely block the window when closed.

Window treatments are also great ways to embellish the shape, style and structural elements of a window. If the window sash and frame are unremarkable, create visual interest with a dramatic sweep of boldly patterned drapes.

If the trim is dynamic and detailed, such as around a Victorian-style double-hung window, let the woodwork speak for itself and use a more restrained window covering, such as blinds installed inside the window well.

THE ROOM-BY-ROOM APPROACH

Although the principles that control how both windows and natural light affect interior spaces are common to any window, each room has its own relationship to light. This relationship is key to deciding how windows will be used in a room.

KITCHEN: A bright and inviting space makes cooking, eating and socializing more pleasurable. If the kitchen receives ambient light, amplify it to prevent the

Strong morning light makes this bedroom a bright and airy space early in the day. Heavy lined drapes make sure that sleeping late is never a problem.

A curved sectional sofa and elegant bow window create a perfect partnership and a stunning backlit centerpiece for this comfortable dining room.

room from appearing overly dark or small. Use gloss paints or finishes on cabinets, and reflective, rather than matte, flooring. Light or bright colors will intensify the light, as will shiny tiled surfaces and chrome or steel accents such as restaurant shelving. If you're fortunate enough to have a kitchen full of windows with strong direct light, too many reflective surfaces can create a cold, harsh clinical atmosphere full of glare and visual hot spots. Where sunlight is abundant, use natural materials to absorb and soften the light. For example, keep food staples in wicker baskets, use satin-finish countertop materials, hang braided strings of garlic and peppers, and incorporate wood accents throughout the kitchen.

One of the best locations for a window is over the kitchen sink. The view provides relief from the tedium of washing dishes and the light is a valuable resource for food preparation. Bay windows are also excellent kitchen additions. They can visually open up a small kitchen and create wonderful cozy spaces for eating breakfast, reading the paper or growing herbs in pots.

LIVING ROOM: The multipurpose nature of a living room means that different areas need to be set up in different ways to take advantage of window exposure. If your living room has a large picture window looking out on a scenic view, exploit it as a major decorative asset. Arrange a seating area in front of the window, much as you would around an entertainment center, to look out on a pleasant view. This is easier in larger living rooms with enough space for several small, self-contained areas, and where placing seating with backs to the common area won't seem so unusual. In more traditional homes with smaller windows, arrange the seating area so that the windows are visual placeholders in the same way wall art is used. An attractive view can be used as a backdrop for a small reading and socializing area. For instance, position the couch between two double-hung windows, using the daylight to frame a comfortable social area.

Function will also help determine where you place living room furniture in relation to windows. A reading chair should be placed near a window with plenty of direct light, while windows that are opposite a TV should be tree-shaded or outfitted with a window treatment that will prevent any glare spots during daytime TV viewing.

Living rooms usually have the wall space to allow for complex window groupings or a stunning focal point window such as a large, divided-lite, arched window with an ornate trim. Modest living rooms are also ideal candidates for corner windows, which will make the space seem larger and can lend detailing

These dining room windows create visual interest with an interesting combination of shapes. Although the window openings are identical, the designer created captivating linear patterns with an asymmetrical design that combines casement and fixed windows on one side, and a row of fixed-glass windows with a casement on the opposite side.

and visual interest to an otherwise boxy layout. Larger living rooms can support a bow or bay window—ideal spaces for a sofa or loveseat.

BEDROOM: At first glance, windows seem to run counter to the primary function of a bedroom—getting a good night's sleep. Use heavy, light-blocking window treatments or position the bed so that the morning light coming through the windows doesn't fall across the bed (or to block urban light sources such as streetlamps). Small fixed-glass windows placed high up can help spread light further into the bedroom, bringing illumination to closets and dressing areas. If the bedroom is large enough to house a separate makeup table or small work desk, you can create a well-lit autonomous area with a bow window, which will leave plenty of room for the desk or table and chair. You should also take into account your preferences for fresh air. For instance, if you like fresh air while you sleep, you will want to position your bed to take advantage of the breeze from an open window.

HOME OFFICE, STUDY OR DEN: Books and reference materials can fade in direct sunlight, so it's often best to position bookcases alongside windows where they'll encounter minimal direct light. A desk placed in the path of direct

▲ An impressive bow window is a simple way to make a spectacular design statement in a basic bathroom. The full-length windows in this apartment bathroom are fitted with bottom-up shades to maintain privacy while allowing light to saturate the room.

▶ Dark wood dining room furniture stands out in this well-lit open floor plan, contrasting white windows and the exceptional lighting exposure. A window placed high up on a facing wall ensures light reaches the dining room and into the kitchen.

light aids in reading or doing paperwork, and a large picture window can ease the feeling of being cooped up in an office. But if most of the work in the office is done on computer, you should position the computer screen so that it doesn't reflect glare from brightly lit windows.

BATHROOMS: A bathroom requires a sensitive balance of privacy and light. But where possible, a large window can make an ordinary bathroom exceptional. Big windows expand the ordinarily restricted space of a bathroom and are an unexpected decorative element. Thoughtful window treatments, such as

Anatomy of a Window

Although the outside frame is perhaps the most obvious decorative element of a window, all the pieces of a window's structure determine its architectural and design style. But modern windows include an additional and essential part of the construction, one that is invisible—insulated glass. Insulated Glass (IG) units are by far the most common types of new windows. IG units are constructed with two panes of glass that sandwich and seal in a layer of inert gas or plain air. Many new windows include a special "low-emission" coating that reduces the loss of heat from the interior. These are designated "Low-E" glass.

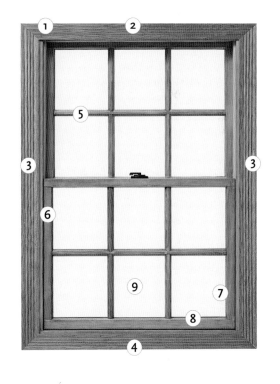

Frame (1): The outside structure that contains the window unit. It is comprised of the **head jamb (2)** on the top, **side jambs (3)** on either side, and **sill (4)** on the bottom. **Muntin/Grille (5):** Muntins are frame pieces that divide separate panes of glass within the window. A mullion is a single structural bar running from top to bottom. Grilles are the modern equivalent of muntins, and are essentially grids that are placed over a single pane of glass to make it look like many different panes, replicating the look of muntins. Unlike muntins, grilles can be removed and replaced to change the look of the window. **Sash (6):** The inner frame that holds glass in place. Traditional windows such as double-hung units are composed of **stiles (7)** (sides) and **rails (8)** (top and bottom). Basic types such as casements have a one-piece sash frame. **Lite (9):** Any single pane of glass, also called "glazing."

▶ An arch-top window creates a spectacular design statement with stylized grilles and a frame that matches the trim in the room in both shape and style. Painting the framing elements white ensures they'll pop out against the red of the wall.

bottom-up shades, can solve the problem of ensuring privacy where larger bathroom windows are used.

Bathrooms with smaller windows should use the bright, reflective surfaces inherent in bathroom design to enhance available natural light. If the bathroom's windows are shaded or out of direct light, consider using bright white fabrics, paint, and chrome accessories. Beige fabrics and paint, and wood accessories, can make a low-light bathroom appear dingy.

Because almost every bathroom has a mirror, you should look to establish a beneficial relationship between the mirror and window or windows. For instance, a mirror opposite a bathroom window that receives bright, direct early morning light, is likely to experience hot spots when you are putting on makeup, shaving or otherwise using the mirror. In this case, you'll want to moderate the light with curtains or blinds, or use another option such as textured or sandblasted window glass to modulate the intensity of the light. Alternatively, you can increase the dim light from shaded bathroom windows by using a larger mirror or adding mirrors.

DINING ROOMS: Windows are seldom the central focus in a dining room, because most formal meals take place at night. But the right positioning can ensure glass-front hutches or glassware sparkle in daylight. Paint the walls light colors to amplify sunlight and draw attention to hanging plate racks and other wall-mounted accents. Use darker colors where you want to diminish the effect

▼ This beautiful arched set of windows is framed by window treatment making the window and view appear as a work of art on the wall. The wood muntins and framing complement the ceiling detail and wood accents.

of the light and make the room seem more intimate. In large or formal dining rooms, complement the room's furniture and trim with archtop windows or windows with decorative grillwork painted a color that contrasts with the walls. This isn't to say that windows can't be used to focus on a view and make meals visual as well as culinary feasts. If the dining room looks out on a garden or other attractive view, and you often dine while it is still light out (or the garden is artificially lit at night) you'll want to ensure a line of sight to the view from as many seats in the dining room as possible.

▶ A simple box frame built out from the windowsill as part of a fireplace base provides a wonderful window seat nook for reading or just relaxing in the afternoon sun. If the fireplace base had been limited to the width of the fireplace, the window would have been underutilized, captured as it was in a recess.

Ventilation is also a concern in some dining rooms, such as when the home is located in a hotter part of the country and there is no air-conditioning. Cross ventilation can effectively cool off a dining room and make for more pleasant summer meals.

THE FRAMING ELEMENTS

The interior trim, sash frame and certain other structural pieces that hold the window and individual panes of glass in place can also serve as decorative accents. In a sense, these help to "trap the view," capturing the scene outside in the context of the home's architecture. The decorative potential of a window's trim elements depends on the interior design and on the home's architectural style. Ornate molding and divided lites will complement a period style such as Victorian, but modern homes call for simpler styles.

Grilles and Trim

Trim elements are the decorative accents that marry windows to a design style. They range from intricately shaped wood pieces to plain metal structures. Simple trim elements will complement many different styles. Or choose windows with more complex trims that blend right in with the home's moldings and woodwork, or a millworked trim for a unique look that draws attention on its own. An untrimmed window is a style in and of itself. In many types of modern and contemporary homes, the sheetrock is run to the edge of the window frame, leaving the window untrimmed. And, in certain situations such as a stone wall, trim elements would be impractical or artificial.

Most commonly, trim styles are dictated by the architectural style of the home. For instance, Tudor and gothic-style windows traditionally feature trim pieces built up of multiple straight pieces, sometimes with accent blocks at the top corners of the window. Period colonial and Federal window frames are simple boxes around the window, accented with rectangular shapes interrupting the trim, and sometimes with a stylized head jamb featuring layered wedge shapes. Victorian window trims are embellished with ornate details, including circles and notches.

But the shape of a window's frame is only one aspect of its decorative potential. Window trim and structural elements can be painted or stained in a full range of finishes to complement or contrast with the wall paint and furnishings. Make the window construction stand out with a bold, high-gloss color, or

43

blend the details with the rest of the décor by using natural clear satin finish or the trim version of the wall color.

Sills and Window Seats

Windowsills are an often untapped decorative asset. Although most sills are simply narrow ledges, they can easily be built out to provide a sun-drenched platform for plants, a showcase shelf for colored glass vases, or as a place for other decorative pieces.

Window seats are the idea of a sill taken one step further. Requiring an ample window recess at least deep enough to accommodate a person sitting comfortably, these are enchanting additions to any house, where they serve as a stage for the sun. Window seats create a well-lit nook suitable for reading, intimate conversations, or simply resting in the warmth of midday. These structures spell comfort and make a wonderful centerpiece of a social enclave shared with friends or immediate family.

Window seats also have their handy side; the underseat area can be used for storage or for display shelves. But the most important potential for a window seat lies in its dramatic presentation. Coupled with a striking window treatment, the window seat becomes a well-appointed partner to the window, creating an undeniably inviting scene.

45

A radiator cover outfitted with custom cushions and comfortable pillows serves as a window seat bench. The window seat provides extra seating in the living room, a bright place to read, and sets a stage for the gridwork and illumination of the windows.

Stained Glass—Painting with Light

If the words "stained glass" bring to mind images of musty cathedrals full of dark and somber tones, think again. The twenty-first century version of stained glass art is as different from its medieval ancestors as a car is from a cart.

In reality, the techniques have changed little. The big difference lies in how and where contemporary stained glass artists use the material. Glass artists long ago realized that the powerful beauty that made illuminated colored glass so compelling in places of worship, would make it an unparalleled addition to other structures private and public, small and large.

Stained glass art can transform an interior space in ways subtle or profound. It can be darkly dramatic, splashing fiery tinted light across a room, or simply a delicate accent that adds spots of subdued hues in unexpected places. The potential range is a result of the diversity in the material itself.

Stained glass is truly stained; the color doesn't rub off and won't fade. Today's

▼ Artist Gordon Huether's *Four Seasons* is a perfect example of how modern stained glass defies the stereotype of a somber medium. These four panels are suspended inside the window wells with thin, almost invisible filament wire. The owners can easily remove the panels if they ever move to a new home.

◀ Combining the privacy of opaque glass with vivid colors and energetic lines, glass artist Samuel Corso's *Framed Landscape VII* brings intensely vibrant hues to an otherwise neutral bathroom design scheme.

stained glass manufacturers offer literally hundreds of colors, color combinations, patterns and textures.The variety ranges from solid colors with uniform quality, to wild swirls and color mixtures with fascinating irregularities. In truth, the stained glass artist works with a palette as varied as a painter's.

But color is just one facet of this unique substance. Stained glasses also vary in the amount of light they transmit, from lightly tinted, nearly transparent designs, to saturated, completely opaque versions. This means an artist can craft a window to suit any architectural, personal or design style. Stained glass can serve anywhere its clear cousin might. Beyond the aesthetic value, the art can play a practical role.

Densely colored opaque glasses can provide privacy while still allowing light to enter. Where insulation value is a concern, the stained glass panel can be sandwiched between two panes of window glass and hermetically sealed, creating a single unit with a higher insulation value than a normal insulated window.

Stained glass windows are also crafted with structural support built in, in the form of the lead channels that hold the individual pieces of glass in place. This allows art glass panels to be installed in any other type of window opening. With such potential in appearance and application, choosing a stained glass window for the home is a process limited only by the boundaries of imagination.

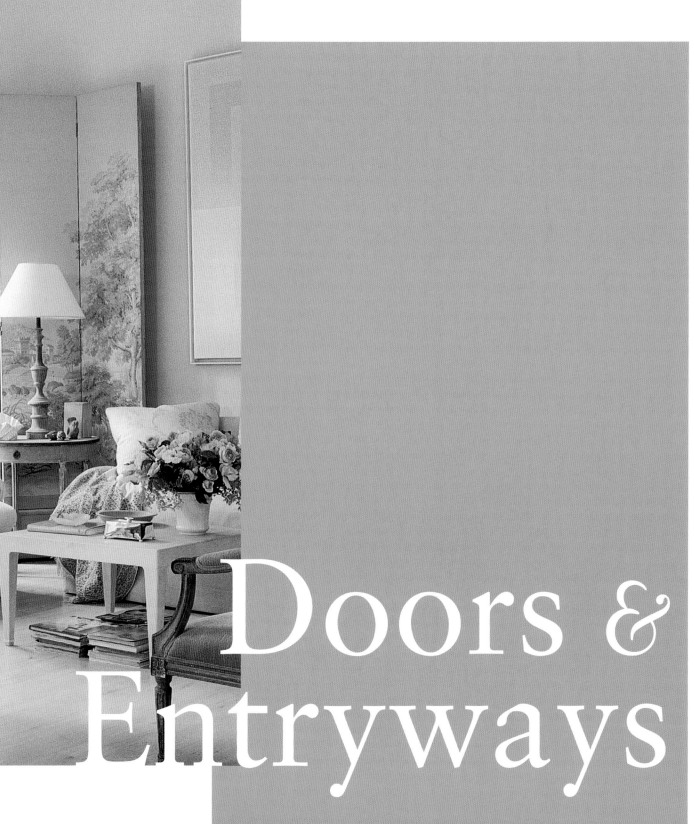

2

Doors & Entryways

▷ Twin French doors serve as central features in this bright and lively kitchen, offering the trifecta of interior design: a fabulous view, tons of light, and essential access to the balcony outside.

Glass doors are delightful visual surprises.

Our minds want to believe that doors are solid barriers, making the clarity and openness of a glass door an unexpected pleasure. Glass in a door also aids the transition between outer and inner space and improves the doorway's graphic appeal, adding a changeable feature that is reflective or transparent depending on available light. To a lesser degree, glass doors also serve the functional purpose windows do—to increase the natural light in a space.

Incorporating glass into doors and entryways is a matter of choosing from an incredibly vast selection of glass and door styles. The choices can be broken down into two basic types of openings: formal entryways and simpler access doors. Entryways are the architecturally designated front entrance and can be as fundamental as a single glass panel in a wood frame, or as complex as a grand arched design full of glass. Access doors, such as mudroom doors or sliding patio doors, are generally less ornate in design, but are essential entry points for light and fresh air, and serve as foils to the garden and the landscaping outside.

Access doors are designed for frequent use and maximum exposure. They usually open onto an attractive outside area, and examples include side, back, patio or deck doors. As functional as they are, access doors present their own design opportunities and are prime candidates for decorative glass treatments.

An entryway carries significantly more design importance. It is the face of your home, the first impression any visitor will have—for better or worse. Entryway style helps define the architecture, whether it's a classic Victorian with a wraparound porch, gingerbread trim and a door with a half panel of stained glass, or a colonial with a restrained solid brink entry and modest sidelites. This is why, if you want the most impact possible with a glass addition, you should start with your entryway.

▼ A daring entryway allows full views in and out of this combination living room and entry space. The door is part of an intriguing geometric design formed by the thick framework. Although it appears as if all privacy is sacrificed, the home actually borders an entry courtyard fronted by a tall stone wall with a solid gate.

Door Types

◁ **Double Hinged:** Larger entryways incorporate double doors hinged on the outside edges, but these are more often used as patio, deck, or garden doors. Double doors are available as in-swing or out-swing to accommodate situations in which the swing space on one side or another is limited. Double doors usually have a locking mechanism in the astragal that holds one door in place while the other is opened.

▷ **Single Hinged:** This is the most common type of door—a simple single door with hinges on one side and a latch and handle on the other. Single-hinged doors suit almost any architectural style. These are also the easiest to install and most useful for basic access doors such as back doors or mudroom entrances. Single-hinged doors can be equipped with full-length or half-length lites.

Swinging: These are sets of two or more doors hinged together in a line. They can be styled to look just like hinged doors, but fold accordion-style out of the way. This allows the doorway to be opened completely. Swinging doors are excellent where interior and exterior areas are regularly combined for entertaining.

Gliding (sliding): Mounted in top and bottom tracks, these work just like gliding windows, with one door sliding horizontally in front of the other. One or both doors of the unit may slide. Sliding doors provide an alternative where any door swing would be a problem. Modern versions are more upscale than their predecessors; manufacturers offer many different grille designs, and you can even purchase French door-style sliding doors. Generally, though, sliding doors are best suited to contemporary homes.

ENTRYWAY OPTIONS

Few architectural features can match the stunning look of a glass-filled entryway. Glass can transform this crucial area from the mundane into the magnificent, changing a space-filler to a design statement. The addition of glass adds sparkle and sheen to the façade of the house, contrasting surrounding textures of wood, brick and stone. Glass panes offer teasing glimpses of the interior, create a sense of anticipation, and fill the entrance with a warm and welcoming glow at night.

Entryway doors differ from their counterparts throughout the rest of the house in that they are generally more complex parts of the architecture. An entryway also frequently features much more than the door itself. It may be enhanced with one or two sidelites—tall, thin openings positioned on one or both sides of the door. More comprehensive designs also incorporate a transom window or windows, positioned above the door.

Entryways don't have to be ornate to be effective. Handsome squares of glass in a transom and sidelites surround a solid door in this entry. The configuration allows for ample light, while maintaining privacy.

The different elements of an entryway allow for a varying design that serves several purposes. The door lite and sidelites in this entryway reinforce the traditional architecture, while the soaring transom provides a stunning focal point to showcase the entrance hall's chandelier.

▲ These door inserts offer elegance and visual interest disproportionate to their size. The Craftsman-style detailing adds a bit of complexity and sophistication to an otherwise plain entryway.

Each one of these elements can stand alone as the glass portion of an entryway. A glass panel in a single hinged door is an excellent way to brighten a basic home façade. Sidelites can add a splash of glass to an otherwise plain entrance with a solid door. A transom window will bring light to the interior and create a bright spot when viewed from outside. However, these individual elements make the most powerful design statement combined in a full entryway design. But whether you are willing to commit to a full entryway of glass, or simply want to include a transparent accent, the process begins with the door itself.

The Door As Design Element

Entryway doors can be a solid foil for the glass in sidelites and transoms or, like the ring for a diamond, they can be the setting for a striking glass installation themselves. Door manufacturers realized this long ago, and now offer as many doors with glass inserts as without. The astounding selection ranges from full-length door panels to small glass portals. Different sizes and shapes of glass "lites" are hardly the only variable. You can select an unfinished door surface, or opt for any of a number of finishes to match existing trim, masonry colors, or other entryway elements. There's also the choice between double and single doors, depending on available space.

The glass itself can vary from unadorned, clear panes, to highly stylized beveled or surface treatments. These days, manufacturers offer all manner of decorative art glass inserts on a mass-produced basis. So while a plain glass door panel is beautiful in its elegant simplicity, you'll find an extensive and tempting selection of art glass inserts to meet just about any taste and design style. Manufacturers also offer a multitude of textured glasses, with the tactile appeal and appearance of pebbles, water, frost and more.

Sidelites: The Supporting Players

The potential for spectacular glass designs in an entryway only begins with the door. Sidelites are a way to complement a glass treatment used in the door, or they can stand alone, providing graphic pop and illumination independent of the door. Sidelites can be wide or thin depending on the site, and can easily fit where there is only a modicum of room alongside the door.

Positioned next to a solid door, sidelites allow the homeowner to maintain privacy while increasing the natural light in a foyer or entry hall. Just like windows, sidelites can be fitted with decorative grids of wood or metal to match other architectural details, or to create a unique look.

In some cases, there may only be room for one sidelite. As long as the combination of sidelite and door is centered within the entryway opening, an asymmetrical pairing will appear visually balanced and can actually increase visual interest. A single sidelite with a divided pane can serve as a bright spot next to a solid wood door with its own raised wood detailing.

Sidelites serve almost as art with elegant Craftsman-style leaded detailing. Notice that although these are decorative sidelites, they fluidly complement all the other glass in the entryway and façade of the house.

Handsome, double-wide transoms add power and light to the French doors in this large living room. The transoms also add scale, balancing the vertical span of glass with the high ceilings.

Topping the Design

An entryway transom window is any glass panel situated above the door. But if you envision a transom as a simple overdoor rectangle of glass, think again. Entryway transoms are available as flowing, intriguing multipane designs. Transoms include an amazing diversity of glass design configurations, from a grouping of small divided lites to a large, impressive arched picture window. They can be a focal point in and of themselves, or the finishing touch on a full-blown entryway. They also serve a valuable practical function, allowing light to penetrate deeper into an entry hall or foyer. Although some entryway transom openings are fitted with operable windows to provide ventilation, in most cases these are fixed-glass windows.

Transoms often serve the same purpose for an entryway that a small accent window does for a high interior wall—that of adding perspective. The transom over a door in a tall entry, such as one that opens into a two-story foyer, brings the doorway into scale with the space.

Transom windows also have all the design potential inherent in sidelites and door panels. They can be contained in frames of many different shapes, from basic squares or rectangles, to flowing curves or polygons. They can also be stylized with decorative grilles and can carry through a design theme established in the door or sidelites.

Complete Systems

Mixing and matching entryway elements can be a fabulous exercise in creativity. It can also be a frustrating challenge if you're not sure of what design direction you want to take with this important architectural space. The easier solution is a premanufactured, all-in-one "entryway system." Complete entryway packages can be the best way to create a completely unified design in which the entryway is seen not as the sum of its parts, but as a single visual piece.

These systems are constructed with sidelites, transom and doors integrated into a self-contained entity. In designing and crafting entryway systems, manufacturers blur the distinction between the different elements. Transoms may meld with sidelites to create a horseshoe of windows around the framework of a door. The doors themselves may be shaped, with arched tops, and transoms that follow the shape.

Whether you prefer the customization potential of combining elements, or the grand possibilities inherent in a system, you'll have to take into account two givens: the physical space and the home's architectural style.

Physical proportions are the biggest constraint on the size and complexity of the glass you install there. Narrow entryways preclude the use of sidelites and mean there is more visual emphasis on the door. Any part of the entrance structure that hangs down may block and diminish the impact of a transom, making it wiser to forgo this element of the entryway.

Architectural style will have a strong effect on entryway design, but it is not as rigid a factor as physical proportions are. On one hand, you can play it safe and let the architecture guide you. Traditional styles such as a Federal, Tudor or Victorian call for smaller inserts—usually less than half the door surface—and less spectacular sidelites and transoms. Full-door glass designs, and picture window transoms are better suited to more streamlined styles such as Prairie School, or modern and contemporary homes. Your home's windows will often lead you in a design direction—if you have smaller double-hung windows, go with smaller panels in the door. If the house includes abundant, plain picture windows, consider an unadorned full-door glass insert.

But the architectural style of your home may not be dominant, in which case you should take more liberties in choosing an entryway design scheme. Where the style of the home is more general or simple—as with many contemporary split-level homes built in the last thirty years—select an entryway that will draw attention to itself and stand out as a principal design element.

As a general rule, undivided clear glass panes are the "beige paint" of architectural details, and will work for every architectural style. Decorative glass panels with ornate flourishes are best suited to homes styles known for that kind of detailing, such as Victorians. Simple linear geometric designs are suitable for a wider range of architectural styles.

The practical needs of the space may also influence entryway glass selection. If your foyer is dimly lit, a full-glass entryway will be the best way to bring light into the space. Even if you decide on a solid wood door, sidelites and a large transom window will greatly increase the amount of light the interior

GLASS FACTS

Fully Formed

Prehung doors are a way to save labor, time and hassle when installing a new door or entryway. These units include the outer framing members into which the door or doors have already been set. Installing them is simply a matter of fixing the entire unit into place. These work best in new openings or in existing door openings that are slightly larger than the prehung unit.

entry hall receives—and this may be the way to go if privacy is a key concern.

The other aspect of lighting is the exterior appearance when the glass is lit from within. Call it the "curb appeal" factor. And though this is a secondary concern to bringing light into the home, in certain cases interior lighting can radically influence entryway design. For instance, a stunning chandelier in a formal foyer begs to be elegantly framed by a large, plain, fixed-glass transom window, showing off the fixture in all its fully lit glory.

A stately colonial entryway is graced by a simple, elegantly beveled half-lite in the door, and half-height sidelites that frame the door in subtle sophistication. Serving as a reflective surface in the shade of the overhang during the day, the entryway is illumined by warm light from inside the foyer at night.

Structural Details

Given how literally flashy entryway glasswork can be, it's easy to underestimate the role of structural detailing, such as the grilles, jambs, and trim. These represent a cornucopia of possible appearances, ranging from the natural warmth of oak, mahogany or cherry, to different stains, to white or colored paint. The structural detailing can complement glass panels in doors, sidelites and transoms, providing visual interest in the form of raised molding, panels, and other fine woodworked details.

Like the overall entryway shape and style, structural details should either complement or purposely contrast with the existing glass, architecture and ornamentation. A modern home defined by lines and angles calls out for an entryway with few or no decorative grilles, and simple trimwork and molding. An Arts & Crafts-style home would be better served with a geometric grille design, perhaps a subdued art glass treatment, and simple square or rectangular shapes in the door, sidelites and transom.

▶ There's no equal to the appeal of a fully lit all-glass entryway. Glass in door, sidelites and the transom reveals the rich natural tones of the woodwork in this entry hall and allows the golden light to wash across the stone and wood surfaces outside the door.

▶▶ Decorative glass in an entryway can also serve functional purposes. The bevels in these Craftsman-style glass inserts refract light, making the most out of the shaded exposure in the entry's recessed position. The glass design itself features bold lines that work well with the chunky steel door handle and the multiple rectangles that frame the door opening.

Inner Space

Modern glass doors—like modern windows—are usually insulated with a layer of air or inert gas sandwiched between two panes of glass. Manufacturers have put that space to good use, suspending shades, blinds or grids in it. The hermetically sealed unit ensures that what's inside never gets dirty or damaged, and exterior controls allow the user to open and shut blinds or lower shades. Manufacturers can also put decorative grids inside the space to mimic the look of true divided lites without the hassle of cleaning and refinishing. The same technology is available in Insulated Glass (IG) windows.

ACCESS DOORS

Glass has traditionally played a role in basic access doors throughout the house. Whether it's a back door with a window that allows for a view of the backyard, or sliding glass doors opening onto a flagstone patio, these types of secondary doors are perfect opportunities to make the most of elegant glass treatments.

The potential for visual impact is greatest where the door opening is largest. A set of sliding glass doors or a pair of French doors are naturally going to be a more significant design element than a single hinged mudroom door, or the back door of a kitchen. That said, there are many ways to make even the most basic access door a glass-filled treat for the eyes.

◀ A double set of French doors adds flair and access to a bright, roomy bathroom. Curtains allow for both privacy and light transmission—essential given the role of the room.

GLASS FACTS

Anatomy of a Door

The double doors shown here are a fairly common type. Except where noted, a single hinged door or a sliding door will have the same structural parts.

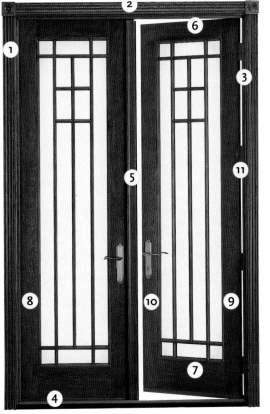

The door frame [1] has the same elements as a window frame: a top piece called **the head [2]**, **the jambs [3]** on either side, and **the sill [4].** (In a sliding door, the sill is actually the foundation piece that contains the rails for the sliding track.) **An astragal [5]** is the centerpiece that keeps one door in place while the other opens—it's not a part of single, sliding or swinging doors. The door itself is comprised of **the top rail [6]**, **bottom rail [7]** (sometimes outfitted with a decorative metal protective plate, called a "kickplate"), and **stiles [8]** (on the hinge side, called the **hinge stile [9]**, and on the handle side called the **lock stile [10]**). The **hinge points [11]** are where the door connects to the frame (these don't exist in a sliding door) and can be concealed, or can be positioned on the outside as a decorative accent.

"Patio" Doors

Manufacturers group any pair of doors not situated in the home's primary entrance under the umbrella term "patio doors." These doors fit in large openings that border backyards, decks, gardens, and yes, patios. Despite the less-than-glamorous title, these can actually be quite elegant and beautiful. They include three different kinds of doors: sliding doors, traditional hinged doors, and folding or "swinging" doors.

Sliding doors are an understated look ideal for tight areas where furniture or other structures would be in the way of door swing. Hinged double doors open out or in and, when completely open, leave the doorway unimpeded by the doors. The same is true of swinging doors, which are hinged at the middle joint and on one edge. They fold out of the way, leaving the opening clear.

At one time, each of these types would have been associated with a different decorative style. Modern door designs have blurred those boundaries and allow the homeowner to marry a desired architectural style with the preferred

◁ Access doors can be integrated into the interior by styling them to fit with a theme. Here, an arched door and frame has been used to carry through the style established by the arched windows in the space.

▽ A pair of patio doors makes for an unusual and attractive addition to a bathroom off a terrace. A moveable screen provides privacy as necessary, while the divided-lite doors allow for maximum light and a terrific view.

If not for the placement of the handle, it would be difficult to tell that these doors are not traditional French doors, but modern gliding versions. With grilles creating the impression of fifteen divided lites in each door, these are every bit as elegant as their hinged counterparts—without requiring the space that swinging doors would need.

opening method. For instance, the classic divided-lite French doors—traditionally double hinged—are now available in both sliding and swinging styles.

Most patio doors incorporate full-length glass panels because they inevitably look out on a garden, backyard or other attractive exterior view. The primary goal is to frame the view, creating a free visual flow between interior and exterior space. The two basic design styles are divided-lite, in which a grille makes the door appear to have many smaller windows, and open-lite forms with as little as possible obstructing the view.

Although view and light exposure generally take precedence with patio doors, the frame, grilles and trim all offer decorative possibilities. Grilles come in many different grids—two-panel, four-panel, six-panel—and, like trim and frame elements, they are sold in many different profiles. All supporting structural elements can be painted, stained or otherwise finished to suit the interior, or better frame the exterior view. A key factor in what type of grids or molding you include with the doors is the architectural style of the home. As with other doors, patio doors are often used to reinforce dominant architectural motifs, echoing the design of windows and other openings.

Undivided panes of glass are a simple and tasteful look, adaptable to most interior and exterior design styles. Generally, the surrounding trim is kept plain and stained natural or painted a neutral color. Because these types of doors allow the scene outside to dominate, they are the style of choice where the landscape is impressive. If you have a showcase flowering garden, or the patio overlooks mountains or a seascape, opt for undivided, full-length glass patio doors.

Although they are not windows per se, both simple undivided glass doors and more complex styles can be enhanced with the right "window treatment." If you prefer the doors themselves to stand out, do without or choose simple vertical blinds or light gauzy curtains to moderate light or establish privacy.

Thick, multicolored drapes grace this pair of French doors and surrounding windows, embellishing the glass openings with all the decadent richness of the furnishings and architectural detailing inside the room.

If you're looking for a more dramatic approach, combine the doors with a striking tumble of drapes and matching valence. Opt for blinds sandwiched between layers of glass in insulated glass doors, or for sash-mounted curtains or blinds, in cases where you want the treatment to not interfere with the movement of the doors.

Single Doors

Single hinged doors are by their nature more modest structures than patio doors. This doesn't mean they have to be unspectacular. The design approach you would take toward an entryway works every bit as well for these more modest openings.

The simplest treatment is a half panel set in an otherwise solid door. The panel can be a simple square or rectangle with edge bevels, or a plain window that does nothing more than allow a view out and light to enter.

The pairing of door panes with a detailed eyebrow transom window ensures that natural light finds its way deep into this confined side entry hall. The trim and grille of the transom complements the other woodwork in the room.

Two single access doors serve as functional parts of a graphic design featuring glass and black frames. The doors serve as parts of a design theme, creating a unified look in this living room.

French doors separating a sun porch from a living room are flanked by wide sidelites that double the visual effect and light transmission of the doors.

Full-length glass doors are a wonderful way to bring a cascade of light into smaller rooms such as the entry hall off a kitchen or a cramped mudroom. Beyond the extra illumination the door brings to the space, a full-length glass door makes small spaces seem larger. It can also bring a clean and bright aspect to the room or area. The one challenge when using these types of door in high-use access areas is keeping the glass clean in a home with children or pets. If you have dogs or small children, you should consider a half panel of glass in the top of the door.

Although they serve the home's design admirably even when simple and undecorated, full-length glass doors can also be as ornate and decorative as entryway doors. As with entryway units, manufacturers offer a variety of full-length art glass panels in single hinged doors.

Access Door Partners

Entryways are not alone in benefiting from the addition of a window above or alongside the door. Any access door becomes a more powerful graphic with the help of an accent window or windows. This can be an extremely appealing way to change the look of a pair of sliding glass doors, or add a unique visual over a single hinged door. Patio doors often have a good deal of wall space alongside them that can be filled with companion windows. The combination of glass doors and awning or picture windows can be an arresting combination, one that optimizes light exposure.

Accent windows can range from two or three transom windows above a wide set of swinging doors, to full-blown arched windows with divided lites and decorative grilles, complementing a set of French doors that open onto a formal garden or flagstone patio.

Style Guide

Hardware Options

Eye-catching hardware can be the ideal finishing touch to any kind of glass door. Door manufacturers and hardware suppliers offer abundant stylish options in handles and hinges. Choose from among a comprehensive selection of finishes—from brushed nickel to traditional brass to powder-coated enamels. The choices multiply with the diversity of handle and hinge styles. You can use a simple knob with a small backing plate that lets the door speak for itself, or select ornate "jug" handles to be a more prominent part of the design scheme. Hinges can be concealed, partially concealed or exposed, in styles ranging from rustic to modern. Let your door style be your guide to choosing hardware that will both complement and enhance the overall glass design.

A Sophisticated Frost

At first glance, sandblasting seems to be a rather limited glass art technique. The treated surface has no color, linework doesn't seem to stand out, and at a distance the designs have little complexity and subtlety.

First glances can be deceiving.

The truth of the matter is that sandblasting is an extremely versatile art form, one that's ripe with potential for creating custom designs on glass doors. Sandblasted compositions can be figural or abstract, geometric or flowing, two-dimensional linear perspectives or three-dimensional shapes. Glass artists can even introduce color to a sandblasted work by painting on the frosted surface or by sandblasting stained glass.

Sandblasting lends itself to photographic accuracy in imagery. The artist applies a "resist" to the glass and then removes sections to be sandblasted in stages. Consequently, almost any design can be sandblasted onto doors—from a favorite photo to letters and words to simple lines and patterns. The edges will appear crisp and well-defined (unless the artist doesn't want them to, in which

▼ Glass artist David Wilson designed a sophisticated geometric motif that was sandblasted across the panels in this modern entryway. The frosted glass surface modulates the light and casts interesting patterns of light and shadow across the interior.

74

Artists Ellen Abbott and Mark Leva created the design for this pair of double doors graced with glass inserts featuring mirror-image sand-blasted designs. The sandblasted design is elegant and complements the formal nature of the entryway, with two frosted stripes echoing the dimensions of the entryway columns.

case the edges can be made to softly dissolve). Using a method called "multi-stage carving," artists can give a sandblasted image subtle shading and amazing visual depth. The process involves sandblasting one area at a time to carve out different parts of the design to different depths. This allows the artist to create a design with the same depth of field a painter might use.

A sandblasted design can serve function as well as form. One of the principal benefits of a frosted surface is that it allows you to maintain a high level of transmitted light without sacrificing privacy. A fully frosted surface is impossible to see through, yet will create a soft diffuse glow in the space, even in ambient light. Sandblasting can be executed on any size pane, making it ideal for doors large and small. The surface itself can be compromised by dirt and grime—such as fingerprint smudges—but is easy to protect with a sealant, or by glazing another piece of clear glass over the sandblasted surface before the panel is set in its opening.

3

Interior Doors & Windows

Second-level interior windows look over a wide-open living room, linking the room with the larger space.

Frosted glass panels in a pair of double doors and in the surrounding wall ensure that this bedroom has necessary privacy while light is generously dispersed between the adjoining rooms. The Lucite panels, glazed with a coat of beeswax, light up with a translucent shimmer even in low-light situations.

It's all too easy to focus on bringing light into the house —

light that you don't know what to do with once it's there. But light travels in a straight line—when it's blocked, it stops. Light won't bend around a wall or find its way through a solid wood door. So the secret to helping natural light reach deeper into the house is to provide pathways through walls and barriers. That's where interior glass doors and windows come in.

These openings provide fuller access to the light, illuminating spaces that are otherwise perpetually dim, and helping adjacent rooms share natural light sources. A strategically placed interior window can almost double the light exposure in a bedroom or filter some of the brightness from a sunlit kitchen into a smaller, darker dining room. A glass door can do the same thing for any room in the house.

As important as it might be, though, dispersing light is but one advantage of glass interior openings. Windows and doors inside the house establish a visual connection between rooms, one that makes both spaces seem a bit larger. Used in this way, the right door or window can tie rooms together fluidly, for example linking a family room with a foyer to give visitors a more welcoming aspect. These installations also embellish the architecture, creating much more visual interest than a plain opening in the same space.

INTERIOR DOORS & WINDOWS

◀ Two doors link two different rooms to this sitting room. Note that although the doors share twin doorways, the doors and transoms are different side to side. The doors on the right contain divided lights, while those on the left use solid glass panels, and the transom on the right incorporates mirrored textured glass, while the transom on the left is filled with clear panes. This provocative design accentuates the visual interest inherent in the shape of the doors.

INTERIOR DOORS & WINDOWS

This visual aspect can be used purposely to create new, interesting and useful views within the house. A special feature in one room can be framed within the window of another. A set of French doors can outline a beautifully laid dining room table or a decorated hearth within the confines of its frame. A glass door or window also lets an adult watch children while preparing food in the kitchen or working in the home office. These types of apertures open the home to inspection so that it's easy to tell who is where at any given moment.

The value of views notwithstanding, any interior window or door has the same graphic potential as an exterior glazed opening. The use of decorative treatments for interior glass applications can have the same impact as art used inside the house.

▶ A rustic living room and adjacent dining room share an antique single hung window that not only helps spread light throughout the rooms, but also serves as a focal point in both rooms. The imperfect glass has ripples that create visual interest in both spaces.

▶▶ The shape and crosshatched muntins of these interesting windows make for eye-catching architectural elements. Simple unadorned windows would have been lost in this elegant interior.

WINDOWS WITHIN

It is simply a matter of construction efficiency that interior windows are rarely designed into a home. It is just easier to plan and build a wall, than a wall with an opening in it. That's an unfortunate reality because interior windows bring so much to a home—enhanced lighting, increased visual range, and remarkable and engaging architectural details that embellish the interior design.

More often than not, interior windows are fixed-glass units. These are less expensive than operable windows, and often the location of the window precludes the access you would need to open the window. Where the window will need to be opened on occasion, gliding types are usually the best choice for interior use. Double-hung windows look a bit odd in an interior, and an open casement window will impede traffic flow in a room.

The Right Site

As with real estate, getting the most out of an interior window is all about location. The spot you select is not just a matter of the lighting exposure, it's also a structural issue. Both sides of the wall should be clear of architectural features that might impede proper installation. You'll also need to take safety into consideration. A window on a wall bordering a family room where kids tend to get rambunctious is an invitation to breakage and the danger that goes along with broken glass.

Certain locations are natural candidates for an interior window. For instance, a wall in an upstairs room that overlooks a large, open downstairs space is an ideal spot for a transparent opening. Rooms that are linked in function, such as a kitchen and dining room or a family room and living room, are also inherently well suited to the use of a glassed aperture. Dimly lit rooms,

◁ Interior windows in a brick wall face their exterior counterparts across a hallway. With the help of a gable-end window and skylights, the light is spread like air throughout the space.

85

GLASS FACTS

Safety's Sake

When using glass in the home, safety should always be the primary consideration. If you have a busy household with kids of all ages and energetic pets, you can choose safer alternatives to standard window glass for interior openings. Tempered glass—the kind used in shower stalls—is much more impact-resistant than regular glass and shatters into small pieces when it does break. You can also opt to install Plexiglas in a fixed-glass opening. In addition to clear types, you'll find smoked or colored styles, or frosted varieties.

such as a bedroom with only one exterior window or a home office with small windows, are also excellent candidates for an interior window.

Transom windows are traditional and extremely useful types of interior windows. Placed over a door or high on a wall, a transom window can spread light into adjacent rooms while maintaining privacy. A functioning transom can also aid airflow and ventilation between rooms. This can be an important issue for rooms such as bathrooms and kitchens, where the opening can help vent humidity and introduce fresh air.

Practical roles aside, transoms and other interior windows add interesting architectural features to an interior design. Make the most of this fact by embellishing the window with details that call attention to it. If the window is used to highlight an attractive feature such as a fireplace or dining room suite, trap the view with a detailed outer frame. For appearances sake, you can use grilles as you would with an exterior window, but grilles or dividers of any sort are usually avoided in interior windows because they obstruct the view from one room to another.

You can also use well-placed interior windows to achieve more subtle—but no less beautiful—aesthetic effects. A window positioned high on a wall between a kitchen and dining room can become a fascinating focal point when backlit by a row of recessed or downlights in the kitchen. This type of situational design is often an excellent way to exploit an interior window.

Window treatments can, in select circumstances, enhance the decorative power of interior windows, but they should be used carefully and sparingly. Curtains or drapes on an interior window can easily seem out of place and even absurd. Where privacy is a concern, the interior window can be made opaque with a glass treatment such as sandblasting, or with the aid of a less obvious window treatment such as shades or shutters. You can also opt to use textured glasses to obscure the view.

▼ Wall-top windows help disperse light throughout this house and provide an open and airy feeling to the rooms on both sides of the wall.

87

When it comes to interior windows, more is often better. Combining them in groups creates an exponentially more powerful interior design graphic, while increasing lighting exposure throughout the space. Let your imagination be your guide. Stack a trio of windows next to a door to mimic the shape of the door and share light between a living room and bedroom. Add a row of small windows high on a kitchen wall to take advantage of the flood of light coming in through a sunroom, or create a collage on a wall between two rooms with several small fixed-glass panes butted together in an asymmetrical layout.

INTERNAL PASSAGES

There are more opportunities for glass doors than windows inside a home because there are naturally more doorways than existing interior window spaces. Most every room with a door is a candidate for a glass door. It's simply a matter of deciding on an appropriate style that goes well with the rooms that are to be linked.

Whatever glass door you choose, it will achieve much the same effect that an interior window does. The door connects two rooms together, facilitating visual harmony in the overall layout of the house. The floorplan of a home with glass doors seems to flow from one space to the next, becoming more a unified space and less a series of boxes. By creating significant exposure to the room beyond, glass doors make each space more inviting.

One of the wonderful aspects of a glass door is that it will block sound while maintaining a sense of connection. Family members can play a raucous board game in a family room while someone reads in an adjacent room—light and the sense of unity are maintained while mutually exclusive activities are accommodated.

Because doors have more surface area than most interior windows, they present great opportunities for decorative line work, such as beautiful grilles or glass art patterns. This means the homeowner in search of a stunning interior glass door faces an embarrassment of riches ranging from a simple solid frame with clear-glass insert, to a traditional divided-lite French door with beveled glass panes.

French by Design

The most traditional and prevalent type of interior glass door is the "French" door—double hinged doors with full-length sections divided into four, six, eight or more lites. Today, manufacturers call any hinged door with divided

◀ An elegant curved transom handsomely complements these double doors, adding a more fluid element to the room and helping light reach even deeper into the interior.

▼ A sleek master bedroom includes identical glass doors on pivots leading to the bathroom and hallway. Versions of the doors have been used for the closet, creating a cool, streamlined aesthetic throughout the room.

lites a French door. This style is widely available and is perhaps the easiest solution to filling a doorway with glass.

The simplest way to fully integrate French doors into a given interior design scheme is to match the number of lites to the complexity of the interior design. For instance, a basic country interior with wide plank floors and comfortable overstuffed furniture is best complemented by a four- or six-pane French door. An interior full of period antiques, elegant molding and fine paintings is better suited to French doors with ten or more lites.

As distinctive as the French door style is, it lends itself to custom treatments that can add even more to the interior. The frame and grilles can be painted to contrast wall color and make an elegant grid stand out against the glass. Or wood grilles can be stained natural to complement other natural wood accents in the room. If the door has separate panes—as opposed to a grille positioned on top of a single pane—you can replace individual lites for a unique effect. Create a checkerboard pattern of stained glass among the clear glass lites, or replace random lites with solid wood panels painted in different colors.

◄◄ Twin pairs of French doors ensure that sounds in the dining room can be blocked off from the living room across the hall, and vice versa. Notice the simple grille in the French doors— basic decorative elements that integrate the doors with the rest of the interior.

◄ A pair of folding French doors create a lovely accent between a formal living room and dining room. These doors are each hinged in the middle to fold back and allow traffic to pass through. The doors themselves serve as an element of the décor, with sophisticated 27-lite panels.

Glass Door Options

French doors may be the most prevalent interior door type, but they are far from the only style. Just as they have with exterior doors, manufacturers have taken the interior door to new design heights, with units that include art glass inserts, decorative trim elements, and different shapes of lites in the doors.

▶ A frosted glass pocket door provides a slick hideaway barrier in keeping with the industrial look of this modern loft.

Style Guide

Riding the Rails

The furnishings in rooms sometimes preclude any door swing. Wall construction in these rooms may also prevent you from installing pocket doors. That doesn't mean you can't install an interior glass door. You can use a special sliding door that runs on rails mounted on the adjacent wall. Although this type of application is not appropriate for all homes and architectural styles, it can be an interesting feature in some contemporary and modern homes.

A line of doors on pivots becomes an unconventional and highly intriguing architectural feature.

Because hinges are the most common way to hang interior doors, most of these decorative glass styles are hinged single doors. However, the pocket door that was common as double doors in studies and formal dining rooms in early-century architectural styles, and widely used as single doors in the kitchens of late fifties and early sixties contemporary ranches, makes for an interesting glass door.

Pocket doors are usually special order items, but can be great showcases for glass. Generally these units are not fitted with grilles, which would interfere with the movement of the door into its "pocket." Instead, they are made decorative with actual divided lights, much like original French doors. They can also be used as canvases for art glass treatments, such as painted glass or sandblasted designs.

Pocket doors can be excellent choices for new construction, providing a sleek, clean look without the challenge of planning for door swing. Although the tracks that pocket doors slide on are unattractive when exposed, a competent carpenter can conceal the tracks.

MAGIC COMBINATIONS

Combining interior doors and windows can make for stunning designs. The most basic type of combination is a single hinged door with a transom above it. Some architectural styles such as Prairie School use this combination throughout a structure as a common theme and a way to keep air moving through the space. The combination can be made lively with the introduction of a geometric art glass design in the transom, door, or both.

More extensive combinations make for attention-grabbing presentations. For instance, a set of French doors leading to a dining room can be supplemented with a border of modest square panes around the outside of the door frame. As simple as it is, this addition carries great visual power and the combination becomes a design element with visual allure exceeding that of either the windows or the doors separately.

◀ These folding French doors not only add an unmistakably elegant passageway between a formal dining room and living room, they also frame an arrangement on the living room fireplace and the sitting area in front of it.

▶ Style doesn't necessarily have to announce itself with a shout. This slick pocket door with frosted glass panels fits right in with the other chic touches throughout the space when closed, and takes up no space when open.

Where wall space is not an issue and you want to make a spectacular entrance to a room, you can use a transom and sidelites or side windows as you might in designing an exterior entryway. For instance, embellish the double doors leading to a formal dining room with a large beveled arched transom window and columns of square beveled panes running top to bottom on either side of the door. This creates an astounding introduction to the room, and stands as its own interior design feature.

◁ Stylish curved-top French doors feature
an eye-catching panel design at the top
that mimics an actual transom.

Art Glass
OPTIONS

Glass artist Larry Zgoda crafted a subtle internal window to complement this living room with a sculpture display area. The muted tones of the textured glass and simplicity of the bevels used add elegance to the space.

A Fine Facet

There is something undeniably elegant about the sparkle of light glinting off the sharp edges of beveled glass. The sparkle is the product of an intentional refined interplay with light, and it is what makes beveled glass art such a wonderful choice for interior use.

The faceted edges of a beveled glass door or window design seem to pick up and enhance the light, even in low-light situations. Through the scientific process of "refraction" (no less beautiful for its logical cause), the beveled edge splits available light into different wavelengths, creating a prismatic effect and sparking transient rainbow colored flashes. The process also creates white highlights that brighten any space.

The lines of a beveled panel are sharp and precise, offering a very clean, crisp look that complements a wide range of interior styles. The glass artist can use ornate curved pieces to make a busy design that fits into a Victorian interior, or she might use straight-line pieces to create a simple geometric design that is better suited for an Arts & Crafts-style home. Whatever the design, the artist will choose between two types of bevels: "stock" or production bevel pieces (called "clusters" and sold in sets), or custom, hand-beveled pieces.

Manufacturers create a dizzying array of shapes and sizes of stock beveled glass pieces. New technology has allowed them to machine-bevel straight pieces and curlicues alike. Artists can mix and match to create wholly original designs for an interior window or door, at a much lower cost than commissioning hand-beveled pieces.

But hand-beveled glass expands the options available to the glass artist. Master bevelers can put a facet on almost any piece of glass that is thick enough to hold it. They can bevel wild shapes, colored glass, and odd bits and pieces such as optical lenses. Hand-beveling also results in slight variations in the angle of the bevels—beautiful imperfections if you will—which make for a greater variety of optical effects when light hits the beveled panel.

Whether stock or custom, a beveled design allows a great deal of light to flow through the space, combining useful transparency that links two adjoining rooms, with enchanting visual effects that provide a constant treat to the eye in light strong and weak, natural and artificial.

▶ Simple doors off a narrow hallway are brought to life with a facet-packed design courtesy of artist Kenneth VonRoenn. The multitude of bevels creates sparkling highlights that amplify and enrich available natural light.

Skylights

The space overhead is the dead zone of interior design.

Although a room may occasionally feature a detailed tray ceiling or a painted ceiling in a distinctive style, by and large we ignore the potential above us. This is usually a case of ignorance rather than neglect. Homeowners and interior designers alike are often at a loss as to how to fill the space in any meaningful way, so they ignore it. But they are ignoring opportunity, the chance to open a room up to the skies and let the light pour in. That opportunity is embodied in a skylight.

A skylight is one of those home improvements that spells luxury in spite of its simplicity. Like other glass installations, skylights expand the dimensions of the house. But with skylights, that expansion is seemingly borderless. With skylights, it really does appear as if the sky's the limit.

And that's to say nothing of the amazing light source these unique openings present. A skylight provides significantly more natural light than a window of the same size. And the light coming in through a skylight penetrates at different angles than the light through windows or doors, complementing the existing exposure and brightening persistent areas of darkness. A skylight will often effectively illuminate a floor space twenty times its own size.

But skylights have another side as well. At night, a skylight becomes a portal to a star-filled sky. A large skylight or a group of them can open up a significant vista of night sky, turning any room into a nighttime platform for stargazing. They become yet another way to link the home with the natural

◄◄ For maximum light it's hard to beat groups of skylights. This sunroom living space enjoys a wealth of natural light courtesy of dual rows of skylights, with ventilating models interspersed among less expensive fixed units.

▼ A living room with modest windows becomes bright and sunfilled with the help of a trio of skylights. The cathedral ceiling precludes the need for skylight "shafts," allowing for direct lighting.

103

▼◄ More than conduits for light, skylights can transform the nature of the space. A perfect example, this windowless kitchen comes to life with the abundant sun it receives throughout the day.

Skylight Types

There are three basic skylight styles. Each has its place in a given room or under particular conditions.

▷ **Fixed Skylights:** By far the most common, these skylights are just that—they don't open. Upscale models are simply flat panels with double insulated glazing similar to insulated windows. Less expensive types are single layers of glass, sometimes with domed or rounded surfaces. A higher price generally means better energy efficiency. Some fixed skylights include a flap that allows for a very modest amount of ventilation.

◁ **Light Tunnels:** These are flexible tubes that bring natural light to normally dark spaces such as closets, and other areas that cannot be fitted with a standard skylight.

Roof Windows: Skylights that open completely and are usually used in situations where the skylight is placed low enough along the roof slope to be accessed and used as a means of egress from the room.

Ventilating Skylights: In addition to bringing in more light, this type of skylight can be opened to increase airflow throughout the space. They may be opened either manually or electrically. Manual versions are often positioned within arm's reach, or within the reach of a specialized turning rod. When opened, a swing arm holds the glass in position, preventing it from opening completely freely. All electric units should be located where they can be wired into the household electrical system.

▲ A pair of skylights brings natural light into a bathroom with a dressing area. Notice how the skylights have been positioned over the vanity area in which the home-owner will put on makeup—natural light is essential for judging appearances.

▶ Three skylights are positioned on the western side of the roof to provide afternoon sun in an attic bedroom. The skylights also provide a view of the night sky.

beauty surrounding it. With all the advantages this type of glass installation offers, the question is no longer what to do with the area overhead, but rather, which room should be the first to have a skylight, how many to install, and what types to use.

PERFECT PLACEMENT

Different types of skylights generally serve different types of rooms. The needs and characteristics of any given room should lead you to a specific skylight, and outside conditions can play a part as well. For instance, if your roof is regularly subjected to strong high winds, you'll want to avoid installing a ventilating skylight because the air pressure variations with the skylight open will lead to slamming doors, flying papers and undue pressure on interior windows.

The primary concerns in choosing a skylight will be light exposure and ventilation. You'll start by determining the existing light exposure and deciding where you want or need additional lighting. If the room has effective ventilation supplied by windows on different walls, along with rotary fans, you probably don't need a ventilating model. If, on the other hand, the room is continually stuffy, it's a good spot for a skylight that opens.

Practical considerations will also play a part in your decision. The more complex or larger the skylight, or the further the ceiling of the room is from the roof of the house, the more involved the installation will be. And installation will be easier with a standard-size skylight rather than a custom model. Most skylights are sized to be easily installed between rafters spaced either 16" or 24" apart, with a minimum of modification. Older, period-style homes may have rafters that are spaced at nonstandard distances.

There's also the question of energy loss. Although skylights are inherently less energy-efficient than windows (air currents are always pushing up against them, with more force than windows encounter), different types vary wildly in energy efficiency. The best units feature insulated thermal glass, which will prevent the unreasonable loss of heat or air-conditioned air. Less effective, but an improvement over plain, single-pane skylights, are those with film coatings that reflect heat away from, or back into, the space. You can also retrofit a plain rectangular or square skylight with a thermal awning or blinds that bounce heat back into the space. The problem with these is that you have to retract them to enjoy the skylight.

> Illumination is no problem in this stylish kitchen; double skylights have been built into the ceiling over the work area. A shaft—essentially a cathedral cutout in the existing ceiling—creates a huge conduit for the light from skylights placed on either side of the roof peak.

Optimize whatever skylight treatment you use by insulating the skylight shaft if one needs to be built for the installation.

Fixed Skylights

Any room with an already established free flow of fresh air is an ideal candidate for a fixed skylight. Installation will be far easier and less expensive than with a ventilating unit, and the light exposure and view will be the same. Because they don't need to be connected to a power source or placed within reach, fixed skylights can be positioned almost anywhere in the ceiling that the physical dimensions will allow.

Fixed skylights create especially powerful graphics when used in groups. For instance, positioning fixed skylights on either plane of a roof will create the impression that a significant portion of the ceiling is actually open to the sky. A trio of fixed skylights along a southwestern side of the room will create the same effect, increasing natural light exponentially. However, whether you're combining them or using just one, the skylight you choose will depend most on the characteristics and the needs of the room.

LARGE OPEN SPACES: Areas such as a combination living room and dining room in an open floor plan most likely already have sufficient airflow. Where interior walls are limited, windows and fans will provide all the ventilation necessary. Fixed skylights greatly increase the projection of natural light into the space, bringing brightness to nooks and corners that aren't illuminated by the windows. A pattern of skylights used across a large open space can be one of the most powerful architectural and design elements you can use. For instance, a living room with high cathedral ceilings can be made even more

Style Guide

Hybrid on High

A roof window is a special type of skylight, part window, part roof opening. Installed in exactly the same way as a skylight is, the roof window looks like a skylight when it's closed. But roof windows can be positioned to open in or out, and they can open completely, in contrast to the partial opening a ventilating skylight features. They are meant to provide potential emergency egress where local codes require one for spaces such as an attic. These units come with most of the same accessories as ventilating skylights, but are not available in electrically operated versions.

impressive with a view of the clouds through facing rows of fixed skylights. It's almost as if you have replaced the roof with glass.

CORNER ROOMS: Rooms with operable windows on two or more walls already have efficient cross-ventilation. Add dual fixed skylights at either end of the room to provide a balanced appearance or use one larger skylight as an overhead focal point. If the corner room is a bedroom, be sure that the skylight exposure will not result in light projecting across the bed early in the morning. Or, if this is the only available position for the skylight given your roof contours, buy a skylight with blinds or a shade built in.

ENTRY AREAS: The flow-through nature of a foyer or entry hall ensures air moves well through the space, but these spaces are often badly in need of

▼ The northeastern exposure in this penthouse living room left the room dark for most of the day. A fixed skylight avoids any problems with high winds and provides abundant light.

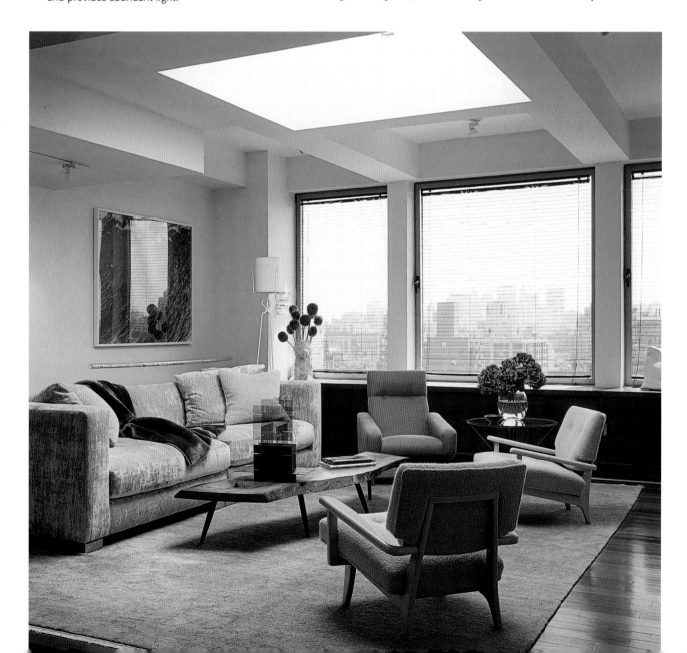

The window seat in this top-floor family room is well-illuminated thanks to a series of skylights that brighten the entire space.

greater illumination. That makes them perfect locations for a basic fixed skylight. The opening will supply significant natural light through the greater part of the day.

A Breeze From Above

Ventilating skylights are wonderful solutions for enclosed spaces, rooms where humidity or vapors of any sort are an issue, and other areas with little or no effective ventilation. Opening like a transom window, even a modest ventilating skylight can provide significant increase in air circulation. These types of skylights create a strong updraft that can quickly clear stale or moisture-laden air out of a room.

StyleGuide

Counterfeit Sky View

You can enjoy much of the appeal of a skylight even in rooms on the bottom floor of a multistory dwelling, or those in which access to a roof-mounted skylight would be otherwise blocked. False skylights can be created in most ceilings by cutting out a shallow cavity and creating a space for special "daylight" bulbs. The cavity is then covered by a layer of semi-opaque glass, such as frosted or heavily textured glass. The effect can be stunning and—unlike the real thing—false skylights provide light at night.

▲ The rustic style A-frame construction of this home is complemented with simple fixed skylights in natural wood frames. Fixed versions are far easier to install.

Ventilating skylights are opened in one of four ways. The simplest types have a crank handle that allows the skylight to be manually opened. Other types are wired into the home's electrical system. These include versions that are opened and closed with a wall switch, those that use a remote control to operate the latch, and automatic skylights that operate by signals sent from a temperature sensor. All of these can be fitted with screens to keep pests out. The one you select will depend on personal preferences and the traits of the room in which you're installing the unit.

A manually operated skylight is most practical where the skylight will be mounted within reach (although some come with a long crank rod that allows you to reach up and open the window). These are the least expensive of the three, but also the least convenient.

Wired-in skylights require the most modifications—especially if there is no existing wall switch into which the skylight can be wired. However, these units are generally the easiest to use. Remote control units are connected to an electrical power supply, but are useless should the remote be misplaced. Automatic skylights are the most convenient; many even have sensors that will close the skylight in the event of inclement weather. But they don't provide the flexibility other types allow you.

ATTICS: Finished attics can naturally seem a bit confined because of severe ceiling slopes, and consequently, air circulation is often meager. A ventilating skylight brightens the space and creates the illusion of more headroom in the process. It can also become a valuable source of fresh air. Use skylights on

Energy Wise

The key factor affecting a skylight's energy efficiency is surface area. Because they usually stick up above a roof's surface, more surface area is exposed than with a window of the same size. Unusual shapes such as plastic domes or bubbles not only contain more surface area through which energy can be lost, but their shapes preclude the use of "Low-e" coatings, insulated glass, and other energy-conserving technologies. Even flat, double-insulated skylights are not as efficient as glass windows or doors because the gas between the panes does not stay as still as in a vertically mounted unit. Newer, more expensive skylights have narrowed the airspace to increase energy efficiency. If you live in a cold part of the country or are concerned about your home's energy efficiency, you'll want to consider one of these pricier units or go the extra expense for operable shutters that close over the skylight and further prevent heat loss at night or during cold periods.

GLASS FACTS

either side of the ceiling to balance the effect and you'll enjoy the added bonus of light throughout the day as well as a multiple-angle view of the sky.

BATHROOMS: The two problems that commonly plague bathrooms are humidity and a lack of direct light. For privacy's sake, windows are generally kept small. In addition, bathrooms are often positioned secondary to other rooms. This means lighting exposure is often less than perfect. A ventilating skylight solves both problems handily. Because any skylight can illuminate many times its own

Twin skylights provide a complement to the walls of windows in this living room. In the open position, the skylights provide effective ventilation by working in tandem with the casement windows below.

A windowless bathroom is an ideal location for a skylight, especially a ventilating model that can quickly clear steam out of the space. This unit has been located right over the shower to give mornings a whole new light.

Three operable skylights efficiently clear odors and stale air out of a top-floor kids room shared by two sisters. The oversized skylights also provide an abundance of light and a view of the sky for the children to enjoy.

surface area, a single unit can often provide an amazing amount of natural light in a bathroom. The updraft created with a door or window open while the skylight is open can quickly remove the steamy aftermath of a long shower. Position the skylight to provide natural light for putting on makeup, shaving or other personal care tasks.

KITCHENS: Light is always welcome and vapors always need to be vented in the kitchen. An opening skylight—or two—can introduce valuable natural light over workspaces and cooking areas, and can bring fresh air into the room, relieving cooking odors. A ventilating skylight also proves a valuable addition over a kitchen table, creating a bright and cheery space for breakfasts or for reading the paper, and will supply a modest breeze on warm mornings.

CRAMPED ROOMS: Closed-off rooms on an easterly side of the house that receive little in the way of cross-breezes may become stuffy and will often have very poor light exposure. A ventilating skylight adds a source of fresh air and creates an effective cross-breeze with any windows or doors in the room, as well as providing much-needed light throughout the day. Use two skylights at either end of the room to brighten it even more, and greatly increase the amount of fresh air.

Sun Censor

Skylights can provide a direct avenue to intense sunlight, but sometimes you need to moderate that intensity. Manufacturers have devised a number of solutions to prevent excess heat and surface fading that can result from a skylight's sun exposure.

Films and coatings are the most basic way to screen out UV and dim intense sunlight. These can be quite effective at blocking out damaging UV rays and moderating sunshine to decrease hot spots off reflective surfaces, as well as preventing fabric from fading. However, the drawback to coatings of any kind is that they cannot be removed when the light is less intense. They can also degrade over time, becoming less attractive and less effective.

For a better option, turn to blinds or shades. For skylights, these are installed in channels that run along the skylight, or in pulley systems that suspend the window coverings in the slanted position below the skylight. Blinds or shades can be electric or manually operated, and come in many different types, ranging from completely light-blocking "blackout" versions, to cellular types that let some light through, to sheer styles that merely diffuse the light. Blinds and shades can also be retrofitted to an existing skylight.

A confined hallway appears brighter and more open with the help of this "sun tunnel."

Pick out clothes in natural light with the help of a light tunnel. This roomy walk-in closet is far more attractive than it would be outfitted just with an artificial light source.

Light Tunnels

Some areas in a home simply won't accommodate a skylight. Rooms with very small ceilings and those that are far below the roof are just not suited to a skylight installation. But that doesn't mean you have to do without natural light; you simply have to use a light tunnel.

Light tunnels are a fairly recent innovation. They are special skylights made of flexible tubes with reflective interior surfaces. The tube actually routes sunlight from the surface of the roof down along a significant length to a round hole in a ceiling. Light tunnels are modest in diameter and, given their flexibility, can be placed through tighter spaces than would allow for a full-size skylight. The ability to be routed through confined spaces allows the light tunnel to direct sunlight into cramped rooms that would usually receive only artificial light.

CLOSETS: A natural light source is a luxury in a closet, whether it's a walk-in clothes closet or a small pantry enclosure. The light tunnel itself is not a significant design feature, but it will introduce a good splash of sunlight, making it easier to see clothing just as it will be seen outside, and helping you find smaller items that might be easily missed in a darker space. And the light from a light tunnel isn't intense enough to fade fabrics, as the direct light from a traditional window might.

SMALL ROOMS: Confined areas such as laundry rooms and mudrooms are made more welcoming and attractive with the illumination from a well-placed light tunnel. Although a small opening, the tunnel can seem to brighten the whole space, making a small room seem cleaner and more vibrant. Rooms dedicated to a given task, such as laundry rooms, are easier to work in with a

Custom Shapes

Although by far the most prevalent skylight shapes are the rectangles or squares most manufacturers produce, you can add a stunning and unique element to your home's architecture and interior design by installing a custom skylight. The many different shapes offer a range of looks and styles (all of these can be manufactured as fixed or operable units, and can be fitted with slots for screens or blinds):

- Slopes or lean-to: This type is shaped like a wedge sitting on its side; the vertical side is butted up to a wall or projection in the roofline. These are handy for roofs with many different intersecting levels, but they should be large enough so that they don't get lost amid the angles of the interior ceiling.
- Ridge: A style that runs along and straddles the crest of a peaked roof. This is one of the best skylights for opening up the ceiling to an impressive sky view and a flood of sunlight.
- Pyramid: As the name implies, the shape projects upward toward the sky. These are most effective when used on flat roofs, where they will maximize sun exposure and provide an extremely interesting visual. A pyramid skylight really comes to life with beveled glass, creating multicolored sparkles that play across the walls and floor below.
- Clusters: Actually a row of square skylights butted end to end to make one long rectangular skylight. This is a great way to replace much of the area of a single roof plane with glass, providing a view from inside that features a wide, unobstructed span of sky.
- Barrel: A vaulted skylight that looks like a barrel cut in half. Often used along the crest line, although it also works just as well on a flat roof.

Custom shapes usually entail a much greater expense than standard prefab skylights, not only because of fabrication, but also due to the more intensive modification of the building that goes into installing the unit. To protect the skylight and moderate the intense sunlight, you may also need a custom plastic cover or other type of sun-blocking device. These skylights can also add expense in terms of energy costs, because the greater surface area increases the potential for heat loss.

pleasant flow of natural light because it's easier to see what you're doing.

HALLWAYS: A sudden burst of natural light in a normally dark hallway space is a welcome surprise, one that can make the hallway seem larger, less narrow and confining. It also goes a long way toward making a more unified floor plan, linking the hallway with other well-lit rooms. You can even incorporate two or more light tunnels in a long hallway.

STYLING SKYLIGHT SHAFTS

If you're fortunate enough to have cathedral ceilings in your home, the skylight will be installed directly into the roof and exposed to the room below. If,

▶▶ The wide and broad skylight shaft in this living room creates maximum light dispersal and adds flair to the room.

▶ A broadly flared skylight shaft ensures that natural light reaches every corner off this bright and inviting bathroom/dressing area.

▽ An octagonal light shaft adds visual interest to this square skylight, and helps disperse light throughout the second floor landing space.

however, the room has a standard ceiling formed by trusses that run across the bottom of rafters, you'll need a shaft to connect the skylight with an opening in the room's ceiling. How that shaft is shaped will affect how the light is dispersed, and will function as an architectural element in and of itself.

A straight shaft is generally dictated by structural conditions. If the space between the roof and the room's ceiling is limited by, for instance, wiring or ductwork, a straight shaft may be the only option.

But in most cases, a flared shaft—one that widens out from the skylight to the ceiling below—is preferable. The first and most important advantage of a flared skylight shaft is that it allows the light to spray out in a wider pattern. This can mean a lot in a room with only one skylight. But a flared opening is also more dramatic and visually interesting than a straight shaft. The flare "opens" the room to the skylight, increasing the visual impact. It also makes it easier to see the view out of the skylight from different areas in the interior.

The Painted Sky

Sometimes it is not enough to have a cascade of sunlight pour down from overhead. Occasionally you may want a more complex effect, full of color and graphic detail. If you feel the room begs for a more dramatic treatment than a simple skylight can provide, turn to the magic of painted glass.

The process of painting on glass creates intriguing gradations of color and shading that make for a magical interplay with light. Skylights are painted in one of two ways. Artists often paint on stained glass with brown or black paints, creating layers that modulate the glass's color and opacity (how much light transmits through), and using heavy black lines to create the outlines or figural work or abstract shapes.

▶ Artist J. Kenneth Leap painted a skylight—entitled *Colonial Skylight*—that not only complements this contemporary interior, but also functions as a unique decorative element on its own. The light modulation qualities of the skylight are more subtle than the stunning visual effect, but no less important.

120

But artists can also paint on clear glass, using special enamels. This allows the artist to control precisely where the color is placed, and usually results in more brilliant and even coloring across the entire design.

Regardless of the method the artist uses, painted skylights come alive when backlit. The saturation of color—in painted stained glass or in enamels on a clear glass—makes for rich jewel tones that delight the eye and draw attention to the painted surface. The color is usually complemented by dramatic linework that adds definition to the design.

These captivating effects can be overpowering, so the skylight needs to be sized and carefully positioned where the impact will be stunning without diminishing the space or other architectural elements. A painted surface can be read in ambient artificial light and is another way the design adds to the space—it just has to be in a position to be seen. The room itself should have other significant sources of natural light to balance the effect of the skylight and supply illumination lost as the paint blocks light transmission.

This close-up shows the incredible detail in the skylight, accounting for its magical appearance. The effect has been created by painting dark paints over stained glass to create highly detailed figural imagery.

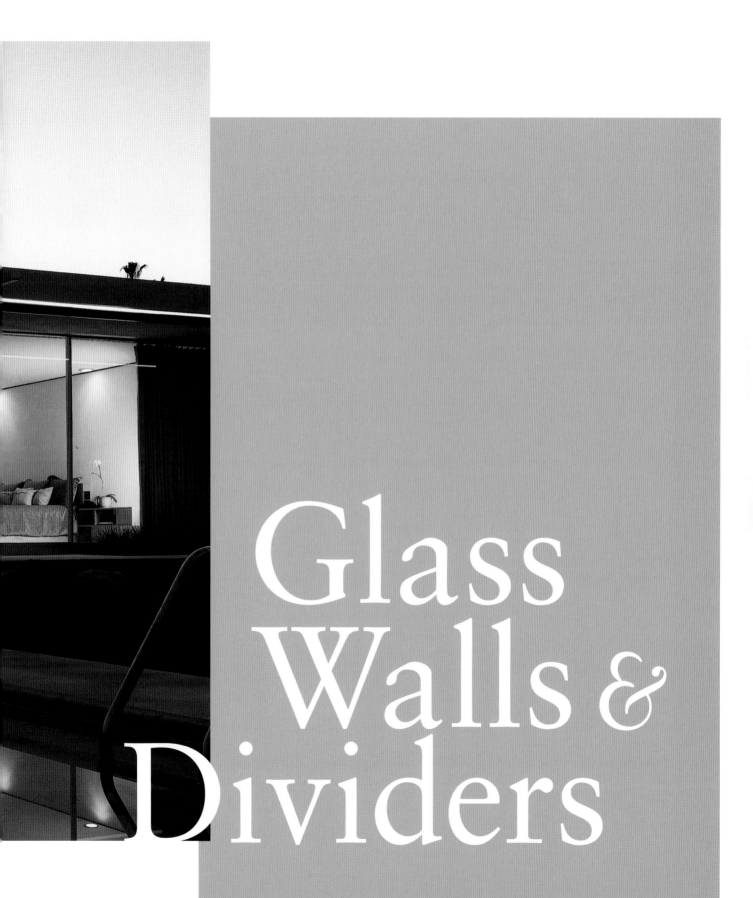

Glass
Walls &
Dividers

For sheer visual power, it's hard to beat a wall of glass.

The marriage of amplified light exposure with the crisp allure of the material itself makes any glass barrier undeniably impressive. Whether it showcases an attractive outdoor area or serves as a transparent separation between areas within the house, a glass wall adds a wholly unique and graphically compelling element to a home's décor.

An exterior glass wall is an unparalleled architectural feature. There's no better way to obscure the boundaries between inside and out, inextricably linking a home with its natural surroundings. You can use a glass wall to highlight a central landscape feature such as garden sculpture, or just make the most of the view from a given room. More importantly, it is a breathtaking conduit for light, brightening the home even on overcast days.

But glass walls also make remarkable interior dividers. An interior wall exploits the dual nature of glass, using it as a physical obstacle while maintaining an open visual pathway. Interior glass partitions allow you to preserve a free-flowing, airy feel throughout a space, while logically separating the floor plan into distinct areas.

Inside or out, the construction of glass walls can be wildly varied, from a simple row of plate-glass windows, to a collage of windows and glass doors, all with thickly profiled frames. Match the glass wall to your architectural style or your own design sense, but the result will be the same: an exceptional home design feature that provides a continual feast for the eyes.

Two stories of glass create an elegant and impressive façade for this modern townhouse. The large spans of plate glass windows are anchored by a row of transom windows, with a casement unit included for ventilation. Two levels of beige drapes provide privacy.

GLASS WALLS & DIVIDERS

A complex wall of glass includes two-story exposure, sophisticated art glass touches in some panels, a suitable entryway that faithfully represents the architecture, and astounding exposure to the light that enriches the interior in any number of ways.

Glass Wall Types

The form a glass wall takes depends not only on the architectural style into which it must fit, but also on the demands of the space and where the wall would be most effective.

△ **Partitions:** A less common usage, these are partial walls used on the inside of the house to provide a break in the flow of the traffic throughout the space or to set one part of a larger area off from the rest. These can be half height or full height, and are often made of alternative glass treatments such as cast glass or glass block.

Exterior walls: Framed as regular walls would be, exterior-facing glass walls can be fitted with large fixed-glass picture windows or can be comprised of several smaller panes contained within their own frames (just be sure the framework does not obscure any part of an attractive view). Exterior walls are often formed from a combination of windows and glass doors, and serve multiple functions including providing a view, ensuring ventilation and allowing for access in and out.

Interior walls: These present more versatile placement opportunities. They can be glazed openings positioned in place of a wall, or can be independent dividers created entirely of framed glass secured between ceiling and floor. Just as with the exterior walls, interior glass walls can include windows and glass doors where the need for a walk-through opening exists.

▶ As beautiful as glass walls can be, controlling the often excessive illumination can be a challenge. Sheer drapes moderate the direct light exposure of this glass wall that borders a stark, modern kitchen and the outdoor sitting area.

TRANSPARENT BORDER

There's just no comparing a blank, unchanging wall with a glass surface that exploits the ever-changing scenery and lighting conditions outside. Exterior-facing glass walls are pure home design power put into play, as different from a static painted surface as fireworks are from confetti.

As wonderful an impact as a glass wall can have, if positioned incorrectly its effect will be muted or even completely diminished. Make the most of this application by positioning the glass wall to overlook a worthy view such as a well-designed garden or fetching copse of trees. A row of floor-to-ceiling windows that faces a side yard with a metal storage shed and garbage cans—or the plain side of your neighbor's house—is a shameful waste of glass.

Light exposure is a somewhat trickier issue. Unlike other glass openings throughout the house, the issue with a glass wall is often how to moderate excessive light. On southerly walls or where the light exposure is particularly strong, you'll need to use some sort of whole-wall window treatment to screen out the sun when it is at its most intense. Vertical blinds are an easy solution, but if you're using double-insulated windows and doors in your wall, you may

The New View

Building a new home is the perfect opportunity to take the idea of glass walls even further. Depending on the surroundings in which your home will be placed, you can use glass walls for a significant portion of the construction—even for all exterior walls. The home should look out on desirable views where glass walls will be placed. You'll also need to consider your privacy requirements. If the landscaping blocks the view into the home, or if

the home is located on a secluded property, you'll have more leeway in the number of glass walls you incorporate. If the house is to be located fairly near to other homes, you'll want to include fewer glass walls, or plan to use privacy screens such as drapes or blinds.

One of the key benefits of a glass wall is that it gives you access to a view. Encompassing one wall of a contemporary home, this glass wall links the stunning desert surroundings with the living room. Double doors help maintain energy efficiency while the air-conditioning is running.

The use of a metal gridwork integrates the door into this glass wall seamlessly. The lines of the framework add visual interest, without lessening the power of the wall itself.

have the option of buying units with blinds positioned inside the sealed units. These pull up out of the way when you want to take full advantage of the view and can block the glass entirely when you need privacy or want to tone down the natural light.

The design style of the home is also going to play a significant part in the decision of where you incorporate a glass wall. For instance, large expanses of pictures windows would look out of place in a Tudor-style home. That's not to say you have to do without the power of a glass wall. A row of floor-to-ceiling leaded panes would suit the architecture just fine and still offer the lighting and design any glass wall should provide. The safest way to create a design for a glass wall is to take cues for framing and glass configuration from existing windows and openings.

A glass wall also needs to match—or purposely contrast—the scope of the property and house. Although the key attraction to a glass wall is how impressive and dynamic it can be, you don't want to make it a sore thumb. This is a matter of exterior view. In a conventional building, such as a Tudor or Victorian structure, replacing a large wall with glass will make the structure look odd. A more modern structure, on the other hand, might be improved with a long run of glass and can even visually support a glass wall across adjacent rooms or up one or more floors, so that a significant portion of the house is actually glass surface.

This townhouse façade makes a powerful statement with a solid red door anchoring three impressive floors of the large plate glass windows.

A window wall takes advantage of a stunning view, providing a seamless link between inside and out.

135

▽ A combination of fiberglass panels, windows and a glass door create an eye-catching and sophisticated intersection of line, translucency and visual texture.

▷ The exterior view illustrates how, with the right framing and positioning, the elements of a glass wall can be neatly integrated with the architecture itself.

Beyond size lies function, and you can build function into the form of a glass wall by adding a door, or by building the glass wall around an existing doorway. Although the most obvious creative direction involves a glass door in the glass wall, incorporating a solid door in a glass wall can be an intriguing reversal of the relationship the mind would naturally expect to see. You can also use a different type of glass for the door, or a decorative treatment such as bevels, to make the door stand out.

THE BLOCK WALL

Glass block is an intriguing alternative to walls of traditional glass. These individual "building blocks" are formed to standard sizes, and tempered to withstand significant construction stresses, imparting strength and durability in window or wall applications. They can be an affordable way to achieve distinctive design elements.

Glass block manufacturing involves rigorous standards, ensuring the dimensions and structural stability are precisely the same block to block. The blocks are hollow, with two sides of matching texture, hermetically sealed with a glass edge. This not only guarantees a measure of safety and stability in a wall constructed of the material, it makes designing and constructing such a wall a simple task. It also ensures a uniform look.

Early on, that uniformity meant visual boredom. The first types of glass block were simple, transparent square cubes of clear, annealed glass. They were more pragmatic than aesthetic, and were usually used as an expedient way to create a physical barrier between two spaces while allowing a maximum of light to filter throughout the space.

Today's glass block styles serve the same purpose, but offer greater design options. Contemporary glass block manufacturers offer exceptional variety,

▼ A simple stepped divider wall provides an interesting graphic shape and sense of separation within the space, creating a serene bathing area perfect for a relaxing soak.

▼▶ This basic grid wall creates a shower enclosure that is at once appealing to the touch and eyes, while resistant to moisture. The transparent wall allows light to penetrate throughout the bathroom.

138

Glass block doesn't need to comprise an entire wall to add an impressive decorative and structural element to the space. The material has been incorporated into the wall of this kitchen and serves as both backsplash and window span for a sunny, bright room. The base of the breakfast island is also constructed of glass block and can be lit from within for an added decorative element.

with a vast number of surface textures ranging from standardized repetitive designs such as raised lines, grids or arcs, to more random textures such as stone, ice or pebble textures. You can also choose from a limited palette of tints, and opacities from clear to completely opaque.

Tinted glass block needs to be used judiciously because the effect borders on neon. It's usually best to use colored glass block as an accent in a much larger wall, or for modest applications such as the base of a jacuzzi tub. You can also opt to install colored fiber optic lights behind the glass blocks—a way to have a splash of color when you want it, and clear block when you don't.

The available textures offer an astounding array of possible looks. In a modern bathroom with sleek lines and stainless steel fixtures, incorporate blocks with ridged surfaces to maintain a minimalist modern feel. A divider wall made of blocks with pebbled surfaces would complement a contemporary kitchen with stone surfaces and natural wood cabinets. You can also mix and match textures for added visual interest. Pick compatible surface textures and create a checkerboard look, or use a small number of contrasting textured blocks to insert a focal point in the middle of a glass block wall.

Physically creating a glass block feature is a fairly easy process. Basically, the blocks are stacked, separated by spacers, then mortared into place. Newly

A second-floor room shares the abundant light of an open common space courtesy of a curved glass block wall. The symmetry of the wall pattern complements the clean and spare lines of the larger room.

An otherwise boring wall bordering a curving staircase is made dramatic with a long span of glass blocks that bathe the steps in light, and sparkle even at night.

introduced silicone bonding with spacers can make installation even easier. The simple construction technique means that a glass block wall can easily be shaped to accommodate function, or to provide a more interesting graphic. For instance, a wall between a dining room and kitchen might include a square opening to serve as a pass-through window. Or create a unique look by making a "stair-step" edge on a wall separating a bedroom from a dressing area. Manufacturers even provide curved glass blocks so that you can create a wavy or curved wall.

As physically rugged as it is simple to work with, glass block makes for wonderful transparent walls in almost any room in the house. Extend a kitchen wall to allow for excellent light dispersion, and a surface that is easy to clean. Install a knee-high divider wall between one level of a home and another, and cap the glass with a wood or stainless steel handrail.

Although it can grace just about any room, the most common application for glass block is in the bathroom. This is a result of the material's inherent properties; the blocks resist both steam and water, and the textures naturally complement the surfaces found in the bathroom. The use of heavily textured glass block also provides essential privacy while maintaining a maximum of light transmission.

Interior Walls and Dividers

Interior glass walls come in one of three forms: a full wall with a door, a complete wall without a door, or a partial partition wall. Whatever its size or extent, an interior glass wall can blur spatial boundaries just as interior windows and doors do, visually expanding rooms within the house.

Although these walls share many obvious features with their exterior-facing counterparts, there are some notable differences. Key among these is the crucial role interior walls play in spreading light—both artificial and natural—

A sliding wall of fiberglass panels opens or closes to allow access to the hallway. It blocks noise and the view when privacy is desired.

throughout the space, rather than just bringing natural light in from the outside. Interior walls also differ in the way they interact with artificial lighting, and the fact that they are often as much decorative elements as conduits for light. Unlike exterior walls, where the view dominates the installation, interior walls are less about seeing through the space than transferring light from one zone to another, or replacing a static wall surface with something more dynamic.

In contrast to a flat, painted wall, glass radically changes character under different lighting. In muted ambient light, a glass divider can disappear, presenting

Simple white wood framing divides this glass wall up into a pleasing composition of rectangular shapes.

the illusion of unfettered space. Lit directly, it sparkles, transmitting and reflecting illumination from a lamp or fixture. This can be a positive if the light source is not too strong. But a high-wattage fixture undimmed by a good light-diffusing shade can create irritating "hot spots" in the glass wall. Consequently, you'll need to avoid placing strong light sources close to a glass wall. The exception is when you use highly textured glasses or glass that has been treated, such as sandblasted panels.

This inset glass wall adds vibrance and light to an otherwise sedate room setting, illuminating the single piece of art in the room.

Frosted glass separates an interior staircase from an adjacent bedroom, allowing enough light to ensure that the narrow staircase is sufficiently illuminated but maintaining privacy in the bedroom.

145

Your choice of glass will also influence how the wall interacts and modifies the light—and view—in the space. Although clear glass is most common, textured glasses can be used to distort the view between rooms and serve as a tactile decorative surface. Simple frosted panels can diffuse harsh, bright spotlights, and offer a pleasant, calming glow when backlit. Using some sort of texture or marking on the glass can also be a safety feature. In certain light, the eye may not register unadorned clear glass walls or partitions, creating the possibility that someone will walk into the glass.

This interplay with the light is just one facet of a glass wall's graphic potential. Framing elements present an opportunity for creativity, allowing you to compartmentalize panes of glass. It's a way to balance positive and negative areas and make use of vibrant linework that captures the eye and intrigues the

▲ A frosted glass partition separates the shower/toilet area from the rest of this bathroom. The glass maintains the sense of the openness and dispersion of light that's a theme throughout this loft, while creating an elegant privacy screen that's easy to clean.

▶ A long glass wall unites a Spartan modern room design with the alluring poolscape beyond.

viewer. For instance, the wall can be made to fit right in with existing elements by bordering the glass with trim molding similar to that used throughout the house, and muntins that echo the home's French doors or double-hung windows. On the other hand, a strong, intentionally unique treatment such as a rugged, powder-coated steel frame will stand out and help make the wall a design statement that purposely contrasts with existing architectural features.

Casting Call

It's a simple fact of life that glass walls are likely to get touched, bumped, whacked and otherwise abused. That's why the ideal glass art treatment for these walls is cast glass. Created from molten glass poured into specially prepared molds, cast glass panels are durable, solid, and as appealing to the fingers as they are to the eyes. The casting process allows a glass artist to create an almost unlimited variety of surface textures. Designs can range from abstract textures such as random striations or brushstrokes, to figural representations. A glass artist can even cast a face or letters into the surface.

Whatever the surface design, light takes on a quicksilver quality when rippling across the textured face of a cast work. The glass itself presents neutral shades evocative of shadows, clouds and fog, with highlights flashing as the wall catches direct lighting. Artists can also cast colored glass, further complicating the changeable optics of the surface. These unique aesthetic effects are combined with an incredibly utilitarian nature. A cast surface is strong and resists breakage, and can serve to maintain privacy where necessary.

▼ A cast entry screen creates a visual break between a large entry foyer and the living room beyond, with a design in relief that provides interest for the eyes and the hands. Glass artist Stephen Knapp created the design in the cast glass especially for the owner of this home.

148

◀ A cast glass partition screen divides one part of a living room from another. Artist Judy Gorsuch Collins individually cast each panel, painting some before firing them. The panels are mounted on a framework of stainless steel supports.

Although the surface reflects and transmits light in captivating and novel ways, the principal attraction of a cast glass wall is often tactile. The surface may be smooth or slightly rough, and touching the wall reveals subtleties that are often hidden from the eye.

In addition to these graphic complexities, panels can be cast with designs on both sides, making the technique an excellent one for interior and exterior walls. Unlike thin brittle panes framed into a wall, cast glass can be several inches thick, with the resilience of a solid stone wall. Where architectural reliability is essential, cast glass is beauty built to last.

Glass
Accents

Glass doesn't have to be grand and impressive

to be a highly influential part of your home décor. Sometimes modest accents, little unexpected touches, can have the same capacity for surprise, intrigue and visual interest as a full-blown glass door, arched window or transparent partition wall. It's all a matter of placing the right accent in the right context.

This isn't difficult, because the material lends itself to myriad applications in rooms throughout the house. The versatility of glass—it can be molded, abraded, painted, drilled, textured and more—makes it a potential replacement for just about any surface in the home, one that combines luminescence, elegance and tactile appeal.

That versatility offers functional advantages. Glass is easy to clean, which is

A cast glass banister insert adds a cool element of sophistication to an otherwise plain architectural feature.

why some of the most natural applications for glass accents include tiles in baths and kitchens, and formed countertops and ledges wherever durability and clean-up are issues. It is also resistant to water and—when properly tempered—is surprisingly tough and resilient, making it the material of choice for shower enclosures and sinks.

The truth of the matter is that there are few places in the home that won't be improved with a touch of glass. It's simply a case of discovering the many options for glass in unexpected places, and then picking the accent that makes the most sense for your décor.

Stained glass cabinet inserts are simple replacements that add a flash of color to this otherwise sleek white kitchen. The glass can be had in a full palette of colors, and is as easy to insert as plain clear glass.

Types of
Glass Accents

Glass accents are all about individual pieces that can be mixed and matched with other materials in the home, and that can be customized to bring a unique look to the room. Most glass accents are modest additions, reasonable in cost but impressive in effect.

◄ **Glass Tiles:** Whether you mix them with traditional ceramic tiles or use them alone, glass tiles add tremendous design impact in kitchens and baths. Available in a full palette of colors, they provide a fresh look to a tiled countertop, wall or floor.

▶ **Fused, Slumped, or Cast Glass Sinks:** Alternatives to the standard porcelain models, they can be textured or smooth, colored or clear. Glass sinks are usually predrilled so that adding hardware and hooking up plumbing are easy to do.

154

▲ **Structural Accents:** Shower enclosures,
cabinet door inserts, and formed counters
are the most common types of structural
accents that involve putting glass to work.

Cast glass tiles make an ideal floor surface, offering a one-of-a-kind look and texture, exceptional durability, and cleanability in one surface.

A fireplace surround and hearth pop out against a neutral interior color scheme when clad in brilliant blue glass tiles. The tiles also make clean-up easy.

HEAD-TURNING TILE STYLE

Glass tiles have become the modern option of choice for homeowners seeking a different and unique look for kitchens or bathrooms. These tiles come in a range of colors, including white and frosted. They also vary from translucent to opaque, but all are as resilient to wear and tear as ceramic tiles are.

Production glass tiles are available in the same standard sizes as ceramic pieces. This allows you to mix and match and easily calculate how many tiles you'll need for a given surface. You can also buy custom glass tiles in unusual sizes, to use either as borders or backsplashes, or to create a one-of-a-kind look on a floor or countertop.

But you don't need to rely on odd-size tiles for impact; a floor, wall or countertop covered completely in glass tile seems to almost glow. The sheen of the surface and the lively colors puts glazed ceramic pieces to shame. You can use a single favorite color to create a powerful, monochromatic appearance, or mix and match in colors to create riveting patterns. Where white is the most common ceramic tile color, the extensive palette of available glass tiles makes bright colors the preferred choice.

As beautiful as they may be in a kitchen backsplash or bathroom wall, the vibrance of glass tiles need not be confined to those rooms. These elegant accents are just as well suited to a well-appointed fireplace surround or as a border around French doors.

Wherever you use them, glass tiles are as easy to install as ceramic versions. The difference lies in preparing the surface underneath the tile. The underlayment for glass tiles should always be painted white and the mastic used to adhere the tile to the surface should also be as bright a white as possible, to illuminate the color of the tiles.

▶ Because of the molten process used to make glass tiles, they can be formed into almost any shape or surface texture. Here, special tiles that look like river stones have been incorporated into a plain tiled bathroom wall, adding an eye-catching splash of style.

▶▶ Glass tile surfaces do not have to be bright and glossy. The sophisticated tiles in this kitchen backsplash offer unique colorations and satin finish that brings a novel look to the wall.

The Mosaic Option

Mosaics are tiny glass tiles, usually less than one inch square, that are used to create fascinating patterns in place of larger tiles. Mosaics are a traditional form of colored glass decoration and in the hands of an artist can be used to make captivating designs on a wall, floor, or countertop. Mosaic tiles are usually intensely saturated with color and tend to gleam like tiny gems. A surface of single color mosaics can be a lovely focal point for a bathroom or kitchen, or you can be more creative and mix several colors in a geometric or flowing design. If you decide on large sections of a single color, you can purchase small sheets of mosaic tiles on prepared webbed backing that allows you to hang a group of them all at once. For a less uniform and generic look, you'll want to deal with individual mosaics. Although more time-consuming, using individual tiles gives you complete control of the design and pattern.

SCINTILLATING SINKS

Traditional porcelain basins are fine for a conservative look, but glass sinks offer an entirely new and fresh perspective on the tried-and-true notion of a bathroom vanity. Tempered for durability, these sinks are as safe and rugged as porcelain or ceramic units, and the incredible selection available ensures that there is one suited to almost any style of architecture and interior design. That selection can be broken down by three factors: color, shape, and texture.

Glass sink colors are as limitless as a painter's palette. Choose one to match or contrast the hue of other fixtures, the tilework, paint on the walls or other design elements. The simplest shades—and the ones that will complement the widest range of bathroom decors—are clear or frosted. (In reality, these usually have a slightly green tint from the iron used in making the glass, although you can buy perfectly clear glass sinks.) For a little variety choose one of the many lightly tinted translucent colors, or go with a more traditional opaque color. Regardless of the color you choose, you can opt to add your own decorative element by having the sink etched with a distinctive pattern.

161

Created using a process known as "slumping," this dramatic sink looks malleable, as if it had been draped over the wrought-iron sink frame. The ability to be crafted in unusual flowing designs is one of the most attractive features about glass sinks.

A cast glass sink and counter complement the glass shower stall and highlight the stark design of this ultra-modern house.

Because of the way glass sinks are formed—melted over a mold or cast in one—they can be shaped into a wild mix of figures, from simple bowls with clamshell ridges, to exotic flowing flower forms. Even if your taste runs to the more conservative end of the scale, you can select among simple square bowls, shaped like trays with a depression in the middle.

Glass sinks are available for the full range of mounting options. Simpler designs of monochromatic opaque colors are great choices as drop-in sinks for vanities. More elaborate and artistic versions can be shown off on a pedestal or in a console frame. You can even buy high glass sinks formed as independent wall-mounted units.

◁ Formed from a single sheet of glass, this sink and countertop fit naturally with the stainless steel base, and the vibrant glass tiles that form a geometric design around the room.

◁◁ A one-piece kiln-formed sink adds a touch of sophistication to an already sleek bathroom. The sink is formed with holes for the drain and faucet.

GLASS ACCENTS

The process of shaping the sink also allows a manufacturer to texture the outside surface. A pebbled or striated surface adds both visual and tactile interest, and can go a long way toward making the sink a focal point of the bathroom. Not only will the sink invite touching and carry an enchanting tactile appeal, light shining through or reflecting off the glass will be broken up into visually interesting highlights.

ENCLOSING THE SHOWER

Where the space is large enough to accommodate it, an all-glass shower enclosure can bring elegance and a luxurious feel to a bathroom's décor. Although these can be exceedingly simple structures, you can also add visual interest by installing an unusual shape—with the added bonus of increasing the shower space. Trim elements, such as the border around the door, door handles and grab bars are

His-and-Her showers complement His-and-Her sinks in this spacious modern bathroom. The large glass shower enclosure contains shower spray without blocking the sense of the room's large dimensions.

A glass shower enclosure becomes a piece of art in this bathroom. Kiln-fired glass panels are painted with abstract black shapes that echo the striations of the marble used throughout the bathroom.

another opportunity to personalize the enclosure with metals that complement other elements in the room.

A glass shower enclosure is also a chance to increase visual flow in the room, and allow greater illumination. A solid-wall shower enclose is a large obstacle in the bathroom; it can block light and can dominate the floorplan of a smaller bathroom. An all-glass version seems nearly invisible, allowing for thorough light transmission and a completely open field of view.

Plain glass walls and doors create a clean and sleek appearance, but if you are looking for a slightly more unique appearance, consider having the glass surfaces lightly etched with a simple design. This is also a way to ensure privacy if more than one person will be using the bathroom at the same time or if small children are likely to barge in. The same effect can be created with cast glass enclosures. It's important to keep in mind, though, if you're considering incorporating a special glass treatment, that you have to make sure the glass is tempered—or clad in tempered glass panels—to prevent injuries in the case of breakage.

▶ An elegant sink is attached to a divider wall of sectioned panes, creating a powerful visual in a lovely bathroom.

▽ Decorative cast-glass handrail sections separate the living room from a second-level bar area in this well-appointed home.

166

Frosted glass cabinets set up an incredible visual interplay with the glass wall against which they've been positioned. The cabinet contents create interesting shapes viewed through the fogged glass fronts, and at night, the cabinet interiors are lit, creating a whole different relationship with the glass wall, which is black.

UNUSUAL SURFACES AND INSERTS

Glass can be at once utilitarian and remarkably beautiful. Nowhere is this more apparent than when it is used in long flat surfaces, or as the front to cabinets and cupboards. These applications are all about function—using glass to replace an otherwise entirely useful surface, and in the process, beautifying a room. Of course, in areas where the glass is subjected to stresses and strains, such as serving as a work surface, the glass must be tempered for safety's sake. Beyond that requirement, the shape, thickness and even color of the surface or insert is a matter of imagination.

Glass can also be incorporated into a surprising number of innovative locations. For instance, used in handrails glass panels can add elegance and sparkle. Thick cast-glass steps can be used in modern staircases for an unusual and visually arresting look.

Counter Punch

Like the best accents in a home, a glass countertop is an unexpected thing of beauty. The lustrous surface is a surprisingly sophisticated alternative to traditional countertops, making Formica, Corian and other ubiquitous materials pale in comparison.

More often than not, glass countertops are suspended above other counters, or out around a structure such as a breakfast bar or island. This allows the glass to feature more prominently in the kitchen's design. But that's not to say that glass can't replace the basic countertop laid over bottom cabinets. The glass can be tinted slightly or made completely opaque with a special film, allowing the material to be used anywhere a counter could go.

Glass countertops can take many forms. They can range from $1/4"$ to more than $2"$ thick, and can be etched with designs, grooved, slumped to create waves, or cast to create a surface in relief. For the most part, however, the basic glass countertop will be standard tempered glass, carrying a slightly green tint.

As attractive as they are, glass countertops are a distinctive style and not ideal for every kitchen décor. For instance, a country kitchen with pine walls and white tilework would be ill-suited to the addition of a glass work surface. These glass accents generally work best in a neutral kitchen that doesn't have a defined style, or more modern and contemporary décors.

Detail of a cast kitchen countertop, showing the interesting irregular surface that has a texture like frosted water.

A stainless steel bar countertop is accented with a simple cast glass surface made by glass artist Dorothy Lenehan. The curved and imperfect surface is a perfect contrast to the sleek steel.

The View Into

Cabinets are wonderful for concealing a variety of foodstuffs and tableware, but often what's kept concealed behind a cabinet door is every bit as attractive as what's left out in plain view. That's why door inserts can be such an effective glass accent; not only do they look crisp and stylish in their own right, they also allow you to show off elegant stemware, fine china, and other beautiful tableware.

Inserts can be plain windows to the interior of cabinets, or they can embody a bit more flair and function as decorative accents. Just as with other "windows" in the house, inserts can be leaded designs with their own tracery of linework in geometric or flowing patterns. They can also be divided into smaller lites occupying the same space, for a more ornate appearance.

But sometimes the view is not desirable, even though glass would add a wonderful decorative touch to the cabinets. In these cases, you can obscure what sits inside a cabinet with a frosted glass insert, or by using a stained glass pane. As with other glass treatments, it's simply a matter of choosing the appropriate options to suit your taste, design style and space.

Special textured glass inserts add flair to this otherwise sedate natural wood cabinetry. The glass has been treated with a "dichroic" coating that makes highlights that change color when viewed from different angles.

Warm Wilting Wonders

One of the many fascinating qualities of a piece of glass is that it can be melted, combined with a different piece of glass, and re-formed to create a new, structurally intact unit. This "fused" glass gives glass artists the ability to meld multiple glass colors in a single composition and form them into any shape for which a kiln mold can be created. Innovative artists have long created dramatic tiles and fused glass surfaces using the technique.

In addition to combining different pieces of glass, the artist can also use ground-up colored glass—called "frit"—much as a painter would use paints. They can produce a figural design or wildly colored abstract patterns. The colors can be sharply defined, one against the other, or can subtly diffuse into each other. The traditional high-gloss surface of a fused tile adds intensity to the colors and makes the tiles themselves visually pop wherever they are positioned.

▶ An eclectic bar area is accented with whimsical fused tiles that glass artist Judy Gorsuch Collins created to capture the fanciful nature of the space.

Vibrant tiles bring this kitchen backsplash to life, with a completely unique look that complements the wood of the cabinets. The tiles are custom-fused in a design created to complement the existing kitchen décor.

The process of fusing involves carefully heating the glass to a precise temperature at which the different pieces of glass will blend but not run like water, and then slowly cooling the glass through stages in a controlled process known as "annealing." This process is critical. It ensures that the structure of the glass is stable, and that any stress is minimized and evenly distributed throughout the glass. Ultimately, this means fused tiles are as durable and break-resistant as ceramic or stone tiles.

Surface effects add to the potential variety in fused tiles. Left alone, the surface will be glossy and smooth. But the artist can fuse the surface over a mold to create a relief, or to just make a random texture. The surface can be painted before fusing to create different shading and variations in color. The artist may even choose to abrade or etch the tile surface to create a rough texture, rather than settle for the typical gloss found on most fused glass tiles.

Because of their intensity, these tiles are generally best used sparingly. Depending on how complex the design and color combinations are, they will be most effective when placed randomly in a tiled wall or countertop of more sedate ceramic tiles in white or neutral colors. However, if your entire kitchen or bath is otherwise monochromatic, you can make a bold statement with a single row, backsplash or entire section of fused glass tiles.

The Language of Glass

A GLOSSARY

ALARM GLASS: Special laminated glass with thin wire in the middle layer. The wire is connected to an alarm circuit, which sets off the security system if the window is broken.

AMBIENT LIGHT: Light filtered into the space from peripheral, nondirect sources.

ANTI-REFLECTIVE GLASS (ANTI-GLARE): Glass with a special coating that reflects very little light.

ARGON GAS: Inert colorless and odorless gas often used in more expensive insulated windows. It is denser than ordinary air, making it a better insulator.

BEAD (GLAZING): Strip of wood, vinyl or other material affixed to or along a sash or sill to hold glass in place.

BITE: A measure of how far the glass extends into the cavity that holds it in place.

BRILLIANT CUTTING: A process by which glass is abraded in a decorative pattern, by use of abrasive and polishing wheels. A common treatment in turn-of-the-century door lites.

BYPASS DOOR: A style of sliding glass door in which the doors pass each other on separate tracks.

CATHEDRAL GLASS: Most common type of stained glass. Usually one uniform color with a texture on one side of the pane.

CENTER SWINGING PATIO DOOR: A patio door with the appearance of French doors, but with hinges between the two door panels, so that only one swings open from the jamb side.

CLERESTORY: Any window positioned on the upper portion of a wall in a room with high ceilings. These are usually placed far out of reach and, when they are operable, are opened with an extension handle.

COLONIAL LITE: The term used to describe fixed or opening windows that are divided into small rectangular lites, and called by the number of lites in the window (e.g., 10-lite, 14-lite).

COTTAGE WINDOW: The term used to describe double hung windows in which the lower sash is larger than the upper sash, sometimes including ornate, decorative muntins.

CRACKLE GLASS: Glass with a distinctive appearance produced by chilling annealing glass so that it cracks, and then sealing it with another structurally sound sheet of glass.

CURB: A custom-built or premanufactured framework that provides the foundation for a curb-mounted skylight.

DOGHOUSE WINDOW: A nonopening window with a peaked top that makes it look like the front opening of a doghouse.

DOUBLESTRENGTH GLASS: Glass that is approximately $1/8$" thick.

ENERGY STAR: A government program sponsored by the Environmental Protection Agency and the Department of Energy, which supplies energy efficiency ratings for windows, skylights and doors, among many other home products.

EXTERIOR BLINDS: Blinds installed on the outside of a skylight, operated by a mechanical pull on the inside, or electrically.

EYEBROW WINDOW: A nonopening window wider than it is tall, with an arched top.

FENESTRATION: Term used for the placement of windows in an architectural wall, specifically describing the graphic placement of the windows.

FLASHED GLASS: A specialty art glass that includes a base coat of one color and a top coat of another color, in the same sheet of glass.

FLAT GLASS: The general term used to describe any sheet of glass originally created in a flat form.

FLOAT GLASS: Flat, clear glass manufactured by floating the molten material on a bed of molten tin in a controlled environment.

GLASS BLOCK: Hollow structural bricks made of tempered glass, and designed to be combined to create a standardized, uniform wall or surface.

GLAZING: The term used to both describe the process of installing loose panes of glass and to describe those panes of glass in place.

GREENHOUSE WINDOW: A window similar to a bow or bay window, but with five sides that project out from an exterior wall. The unit is framed as a single piece, often with shelves for holding plants.

HEAT GAIN: The process by which heat coming through a glass opening collects on the inside of a building.

HEAT-TREATED: A term that is sometimes used to describe tempered or heat-strengthened glass.

HOPPER: A type of window that is hinged on the sides or bottom, opening at the top by swinging in or out.

HORIZONTAL ROLLER: An alternate term used for a sliding or gliding window.

INTERNAL MUNTINS: Grids contained in the cavity of a double-insulated window or door, meant to create the look of divided lites.

LAMINATED GLASS: Different from tempered or insulated glass, laminated panels are constructed of two layers of heat-strengthened glass sandwiching a layer of vinyl that will hold the glass in place if it is broken.

LIGHT SHAFT: The column that connects a roof-mounted skylight to the hole in the ceiling of a room below.

LOW-E (LOW-EMISSIVITY): A coating on the surface of the glass that prevents heat transfer and transmission of UV rays.

MACHINE GLASS: Glass with a uniform texture that is created during a process that involves a mechanized mold rolling over molten glass. as it cools

ORIEL WINDOW: A type of bay window that projects out from a building, and is usually supported from below by corbels or brackets.

PASS-THRU WINDOW: A single-hung window used in interiors as a functional channel between rooms such as a kitchen and dining room. Usually placed over a counter.

PATTERNED GLASS: A type of textured glass impressed with uniform patterns.

PLATE GLASS: Traditional window glass made by rolling or casting—highly breakable and used rarely in new applications.

PROJECTED WINDOW: Any window that opens outward, such as a casement or awning.

R-VALUE: A measure of a glass surface's resistance to heat transfer. The higher the R-Value, the better the insulating ability of the glass.

ROLLED GLASS: Sheet glass manufactured by passing molten glass through rollers to make a sheet with exact thickness and surface patterns. Glass can be rolled by machine or by hand.

SAFETY GLASS: A type of laminated glass that has passed strict impact tests for resistance to breakage under pressure.

SELF-FLASHING: A type of skylight that creates a watertight seal when installed, without the need for other flashing or protective structures.

SINGLE-STRENGTH GLASS: Sheet glass that is approximately $3/32"$.

STACK: Term used to describe two or more windows positioned one above the other. Also, a window or windows positioned directly above a glass door.

TEMPERED GLASS: Hardened glass that has better resistance to breaking than standard window glass, and will shatter into small, uniform pieces rather than shards when broken.

U-VALUE (ALSO U-FACTOR): The measure of heat flow from inside a structure, through glass, to the outside. The lower the U-Value, the better the insulating ability of the glass.

UV: Ultraviolet light. UV rays can fade fabrics and, in large amounts, can be harmful to skin. Most glasses absorb a certain amount of UV light, but many manufacturers offer coatings for window and door glass, which block out UV rays entirely.

WIRE-REINFORCED GLASS: Sheet glass that has wire mesh laminated between the layers. Used where there is a greater-than-normal risk of breakage.

Photography Credits

Page 1: Peter Margonelli

Page 2: Minh + Wass

Page 3: Jean-Francois Jaussaud

Page 5: Peter Aaron, Alexander Gorlin Architect

Page 6: Carlos Domenech

Page 7: Courtesy of Pella Windows & Doors, www.pella.com

Page 9: Grey Crawford

Page 11: Peter Aaron, Alexander Gorlin Architect

Page 12: Courtesy of Oceanside Glasstile, (760) 929-4000, www.glasstile.com, photo by Christopher Ray Photography

Page 13: William Waldron

Page 14: Tim Street-Porter

Page 15: Oberto Gili

Page 16: Timothy Hursley

Page 18: Oberto Gili

Page 19: William Waldron

Page 20 top left: Eric Piasecki

Page 20 top right: Laura Resen

Page 20 bottom: Tria Giovan

Page 21 top left: Courtesy of Crestline Windows and Doors, www.crestlinewindows.com

Page 21 top right: Tria Giovan

Page 21 middle: J. Savage Gibson

Page 21 bottom: Courtesy of Crestline Windows and Doors, www.crestlinewindows.com

Page 23: Eric Piasecki

Page 25 both: Jeff McNamara

Page 26: David Glomb and Julius Shulman

Page 27: Oberto Gili

Page 28: William Waldron

Page 30: William Waldron

Page 31: Tria Giovan

Page 32: Oberto Gili

Page 33: Courtesy of Pella Windows & Doors, www.pella.com

Page 34: Grey Crawford

Page 35: Gordon Beall

Page 36: Thibault Jeanson

Page 37: Eric Boman

Page 38 top: Peter Margonelli

Page 38 bottom: William Waldron

Page 39: Courtesy of Pella Windows & Doors, www.pella.com

Page 40: Tria Giovan

Page 41: Tim Street-Porter

Page 42: Peter Margonelli

Page 44: Laura Resen

Page 46: Misha Bruk, design by Gordon Huether

Page 47: ©Chipper Hatter, Hatter Photographics; artist Samuel Corso

Page 48: Gordon Beall

Page 50: Oberto Gili

Page 51: Peter Aaron, Alexander Gorlin Architects

Page 52 top: Courtesy Cast Glass Images, Inc., www.castglassimages.com

Page 52 bottom: Courtesy of Pella Windows &
Doors, www.pella.com

Page 53 top: Courtesy of Peachtree Doors and
Windows, © 2006 Peachtree Doors and Windows,
www.peachtreedoor.com, 800 732-2499

Page 53 bottom: Courtesy of Therma-Tru Doors,
www.thermatru.com, 1-800-THERMATRU

Page 54: Oberto Gili

Page 55: Courtesy of Therma-Tru Doors,
www.thermatru.com, 1-800-THERMATRU

Page 56: Courtesy of Therma-Tru Doors,
www.thermatru.com, 1-800-THERMATRU

Page 57: Courtesy of Andersen Windows, Inc. ©
2006 Andersen Corporation,
www.andersenwindows.com, (800) 426-4261

Page 59: Gaby Zimmermann

Page 61: J. Savage Gibson

Page 62: Courtesy of Pella Windows & Doors,
www.pella.com

Page 63: Courtesy of Signamark, © 2006 Trinity
Glass, www.signamark.com

Page 64: Tim Street-Porter

Page 65: Courtesy of Pella Windows & Doors,
www.pella.com

Page 66: Gordon Beall

Page 67: Jacques Dirand

Page 68: Courtesy of Peachtree Doors and Windows,
© 2006 Peachtree Doors and Windows,
www.peachtreedoor.com, 800 732-2499

Page 69: Minh + Wass

Page 70: Dominique Vorillon

Page 71: Simon Upton

Page 73: Jeff McNamara

Page 74: Richard Walker; design by David Wilson of
David Wilson Design, South New Berlin, New
York; glass etching by Denise Leone

Page 75: Ellen Abbott and Marc Leva, photo and
glass design

Page 76: Peter Aaron, Alexander Gorlin Architects

Page 78: Courtesy of Andersen Windows, Inc. ©
2006 Andersen Corporation,
www.andersenwindows.com, (800) 426-4261

Page 79: Oberto Gili

Page 80: Jacques Dirand

Page 82: Evan Sklar

Page 83: Gordon Beall

Page 84: Courtesy of Andersen Windows, Inc. ©
2006 Andersen Corporation,
www.andersenwindows.com, (800) 426-4261

Page 86: Courtesy of Pella Windows & Doors,

www.pella.com

Page 87: Grey Crawford

Page 88: Jacques Dirand

Page 89: Richard Bryant/Arcaid

Page 90: Gordon Beall

Page 91: Gordon Beall

Page 92: Christopher Wesnofske; Belmont Freeman
Architects

Page 93: Peter Aaron/Esto, Alexander Gorlin
Architect

Page 94: Tria Giovan

Page 95: Victoria Pearson

Page 96: Tim Street-Porter

Page 98: Larry Zgoda, photo and design

Page 99: Kenneth VonRoenn

Page 100: Courtesy of Andersen Windows, Inc. ©
2006 Andersen Corporation,
www.andersenwindows.com, (800) 426-4261

Page 102: Courtesy of Andersen Windows, Inc. ©
2006 Andersen Corporation,
www.andersenwindows.com, (800) 426-4261

Page 103 both: Courtesy of Velux America, Inc., 800-
888-3589, www.veluxusa.com

Page 104 left: Courtesy of Velux America, Inc., 800-
888-3589, www.veluxusa.com

Page 104 right: Courtesy of Velux America, Inc., 800-
888-3589, www.veluxusa.com

Page 105 top: Courtesy of Velux America, Inc., 800-
888-3589, www.veluxusa.com

Page 105 bottom: Courtesy of Velux America, Inc.,
800-888-3589, www.veluxusa.com

Page 106: Courtesy of Velux America, Inc., 800-888-
3589, www.veluxusa.com

Page 107: Courtesy of Andersen Windows, Inc. ©
2006 Andersen Corporation,
www.andersenwindows.com, (800) 426-4261

Page 109: Eric Piasecki

Page 110: Roger Davies

Page 111: Courtesy of Velux America, Inc., 800-888-
3589, www.veluxusa.com

Page 112: Courtesy of Velux America, Inc., 800-888-
3589, www.veluxusa.com

Page 113: Courtesy TRG Architects,
www.trgarch.com

Page 114 both: Courtesy of Velux America, Inc., 800-
888-3589, www.veluxusa.com

Page 115 both: Courtesy of Velux America, Inc., 800-
888-3589, www.veluxusa.com

Page 116 left: Courtesy of Velux America, Inc., 800-
888-3589, www.veluxusa.com

Page 116 right: Courtesy of Velux America, Inc., 800-888-3589, www.veluxusa.com

Page 118 both: Courtesy of Velux America, Inc., 800-888-3589, www.veluxusa.com

Page 119: Courtesy of Velux America, Inc., 800-888-3589, www.veluxusa.com

Page 120: J. Kenneth Leap, photo and design

Page 121: J. Kenneth Leap, photo and design

Page 122: David Glomb and Julius Shulman

Page 125: Peter Aaron, Alexander Gorlin Architects

Page 126: Courtesy of Andersen Windows, Inc. © 2006 Andersen Corporation, www.andersenwindows.com, (800) 426-4261

Page 128: Oberto Gili

Page 129 top: Timothy Hursley

Page 129 bottom: Peter Aaron, Alexander Gorlin Architect

Page 130: Grazia Branco

Page 131: Peter Aaron, Alexander Gorlin Architects

Page 132: Dominique Vorillon

Page 133: Tria Giovan

Page 134: Courtesy of Alexander Gorlin Architects

Page 135: Courtesy of Andersen Windows, Inc. © 2006 Andersen Corporation, www.andersenwindows.com, (800) 426-4261

Page 136: Anthony Cotsifas

Page 137: Anthony Cotsifas

Page 138 left: Courtesy of Pittsburgh Corning Corporation and Seattle Glass Block

Page 138 right: Peter Aaron/Esto

Page 139: Courtesy of Pittsburgh Corning Corporation

Page 140: Courtesy of Pittsburgh Corning Corporation

Page 141: Courtesy of Pittsburgh Corning Corporation and Seattle Glass Block

Page 142: Anthony Cotsifas

Page 143: Peter Aaron, Alexander Gorlin Architects

Page 144: House Beautiful

Page 145: Christopher Wesnofske; Belmont Freeman Architects

Page 146: Christopher Wesnofske; Belmont Freeman Architects

Page 147: House Beautiful

Page 148: Courtesy of Stephen Knapp

Page 149: Ron Johnson; artist: Judy Gorsuch Collins

Page 150: Courtesy of Oceanside Glasstile, (760) 929-4000, www.glasstile.com, photo by Christopher Ray Photography

Page 152: Courtesy of Seattle Stained Glass

Page 153: Simon Upton

Page 154 top: Courtesy of Interstyle Ceramic & Glass Ltd., www.interstyle.bc.ca, (604) 421-7229

Page 154 bottom: Courtesy of Oceanside Glasstile, (760) 929-4000, www.glasstile.com, photo by Christopher Ray Photography

Page 155: Peter Aaron, Alexander Gorlin Architect

Page 156: Courtesy of Interstyle Ceramic & Glass Ltd., www.interstyle.bc.ca, (604) 421-7229

Page 157: Courtesy of Oceanside Glasstile, (760) 929-4000, www.glasstile.com, photo by Christopher Ray Photography

Page 158: Courtesy of Interstyle Ceramic & Glass Ltd., www.interstyle.bc.ca, (604) 421-7229

Page 159: Courtesy of Interstyle Ceramic & Glass Ltd., www.interstyle.bc.ca, (604) 421-7229

Page 160: Photo by Bill Timmerman. Architecture: House of Earth and Light designed by Marwan Al-Sayed. Cast Glass Sinks designed by Mies Grybaitis and Marwan Al-Sayed. Courtesy of Marwan Al-Sayed Architects Ltd, www.masastudio.com

Page 161: Courtesy of American Standard, www.americanstandard-us.com

Page 162: Courtesy of Lacava Bathroom Design, www.lacava.com, (888) 522-2823

Page 163: Courtesy of Interstyle Ceramic & Glass Ltd., www.interstyle.bc.ca, (604) 421-7229

Page 164: Courtesy of Lacava Bathroom Design, www.lacava.com, (888) 522-2823

Page 165: Kenneth VonRoenn, photo and design

Page 166: Courtesy of Seattle Stained Glass

Page 167: Eric Boman

Page 169: Peter Aaron, Alexander Gorlin Architect

Page 170: Dorothy Lenehan

Page 171: Courtesy of AFG Glass, www.afgglass.com

Page 172: Ron Johnson; artist: Judy Gorsuch Collins

Page 173: Ron Johnson; artist: Judy Gorsuch Collins

Page 174: Gordon Beall

Page 180: Tria Giovan

Page 184: Dominique Vorillon

Index

Boldface page references indicate primary discussions.

Underscored references indicate boxed text. Italic references indicate illustrations.

A

Accents. *See* Glass accents
Access doors. *See* Doors (access)
Art glass
    beveled, 98–99
    cast glass, 148–149
    in doors, 62, 63, 74–75, 98–99
    fused glass, 172–173
    in glass accents, 172–173
    in glass walls, 126–127, 148–149
    inside house, 98–99
    intense colors, 172
    painted finishes, 120–121
    sandblasted glass, 74–75
    in skylights, 120–121
    stained glass, 46–47
    in windows, 46–47, 98–99
Attic skylights, 112–113
Awning windows, 21

B

Banister inserts, 152, 166
Bathroom

access doors, 64, 65, 67
block walls, 138, 140
frosted glass wall in, 146
glass sinks, 154, 161–165
glass tiles, 154, 158, 163
shower enclosures, 155, 164,
  165–167
skylights, 106, 113–114
windows, 39–41
window treatments, 34
Bay windows, 21
Bedroom
    block walls, 140
    skylights, 106, 107
    windows, 28, 29, 35, 38
Beveled glass, 98–99
Block walls, 138–141
Bow windows, 21, 36, 37, 38

C

Cabinets/inserts, 153, 168, 169, 171,
  173
Casement windows, 20, 85

Cast glass, 148–149

Cleaning windows/glass, 29

Closet skylights, 116–117

Color, of light, 24, 26–29

Countertops, 168, 170, 171

Cross-ventilation, 19, 85

D

Den windows, 38–39

Dining room

    French doors, 90, 91

    skylights, 108–110

    windows, 36, 38

Directional differences, 12–15

Doors (access), 50, 65–73

    accents complementing, 72–73

    anatomy of, 64

    art glass options, 74–75

    hardware options, 72

    "patio" doors, 66, 67–70

    single hinged, 70–72

Doors (entry). *See also* Entryways

    anatomy of, 65

    art glass in, 62, 63, 74–75

    blinds/shades inside glass, 64

    complete systems, 58–61

    curb appeal factor, 61

    as design elements, 56

    in glass walls, 136, 137

    hardware options, 72

    overview, 50

    prehung, 60

    sidelites with, 54, 55, 56–57

    transoms with. *See* Transom windows

    uniqueness of, 55

Doors (interior), 79–82, 89–99

    anatomy of, 65

    art glass options, 98–99

    benefits of, 79–82, 89

    combining windows with, 94–97

    decorative features, 89

    divided lights, 81

    French, 88, 89–91, 94

    glass, options, 92–93

    pivoting, 93

    pocket, 92, 93, 94, 95

    rail-mounted, 92

    solid lights, 80–81

Door types

    double hinged, 52, 67

    gliding (sliding), 53, 67, 68, 92

    single hinged, 52, 70–72

    swinging, 53, 67

Double hinged doors, 52, 67

Double-hung windows, 20, 26, 27, 30

E

Entryways. *See also* Doors (entry)

    architectural style and, 60

    complete systems, 58–61

    curb appeal factor, 61

    importance of, 51

    options, 55–63

    practical considerations, 60–61

    simple elegance, 54, 55

    skylights in, 110–111

    structural details, 62

Exposures, of light, 15, 24–29

    eastern, 26

    environment and, 26–29

    most desirable, 26

    nighttime, 26

    redecorating existing rooms and, 30

    southern, 26

    southwestern, 25, 26

    western, 28, 29

    working with existing windows, 30

F

False skylights, 111

Fenestration. *See* Window

placement/fenestration

Fireplace surround, 156, 157

Fixed glass ("picture windows"), 20, 25, 26, 27

Fixed skylights, 104, 108–111

Floors, glass tile, 156

Fogging panels, electric, 144

French doors
  glass walls complementing, 146
  gliding (sliding), 68
  inside house, 88, 89–91, 94
  interior windows complementing, 87
  kitchen entry, 50
  living room to patio, 72, 73
  transoms above, 58, 59, 88, 89, 96, 97
  treatments complementing, 69, 70, 91

Frosted glass walls, 144, 145, 146

Fused glass, 172–173

G

Glass
  benefits of, 12–15
  safety considerations, 85
  versatility of, 152–153

Glass accents, 150–173
  art glass options, 172–173
  cabinets/inserts, 153, 168, 169, 171, 173
  countertops, 168, 170, 171
  fireplace surround, 156, 157
  flooring, 146
  overview, 152–153
  shower enclosures, 155, 164, 165–167
  sinks, 154, 160, 161–165
  structural accents, 155
  tiles, 154, 156–159, 163, 173
  types of, 154–155
  unusual surfaces/inserts, 168–171
  versatility of glass and, 152–153

Glass walls
  art glass in, 126–127, 148–149
  in bathrooms, 138, 140, 146
  in bedrooms, 140
  benefits of, 124, 132
  block walls, 138–141
  bordering staircases, 140–141, 145
  exterior walls, 129, 130–137
  framing elements, 143, 145–146
  frosted glass options, 144, 145, 146
  functionality of, 136
  in hallways, 142
  home design/style and, 132–135, 136, 137
  interior walls/partitions, 124, 128, 129, 138–141, 142–147, 148–149
  in kitchens, 130, 131, 139, 140
  light exposure/control, 130–132
  in living rooms, 132, 135
  new home plans and, 130
  outside surroundings and, 124, 130, 132
  overview, 124
  privacy considerations, 130, 144
  selecting glass type, 145
  self-fogging panels for, 144
  two-story, 124, 125, 126–127
  types of, 128–129
  varied materials/textures, 136, 139, 140
  views through, 130, 132, 135
  visual power of, 124
  window treatments for, 130–132, 144

Gliding (sliding) doors, 53, 67, 68, 92

Gliding (sliding) windows, 21

Glossary of terms, 174–178

H

Hallway
  glass walls in, 142
  skylights, 111, 116, 118, 120

Handrails, 152, 166
Hardware, for doors, 72
Home office windows, 38–39

K

Kitchen
   block walls in, 139, 140
   cabinets/inserts, 153, 168, 169, 171,
      173
   entry doors, 50
   glass tiles in, 156–159
   glass walls, 130, 131, 139, 140
   skylights, 108, 109, 114
   tile backsplash, 158
   windows, 35–36

L

Laundry room skylights, 117–118
Light (natural)
   colors of, 24, 26
   direct vs. indirect (ambient), 24
   environment coloring, 24, 26–29
   exposures, 15, 24–29
   importance of, 12–15
   paint/finishes and, 25, 29
   penetrating into house, 79
   producing vitamin D, 12
   SAD and, 12
Light tunnels, 104, 116–118
Living room
   access doors, 70, 71
   entry doors, 51
   glass walls, 132, 135
   interior windows in, 78, 79, 82, 83
   skylights, 108–110, 112, 113, 118,
      119
   windows, 32, 33, 36–38

M

Mosaics, 159
Mudroom skylights, 117

N

Natural light. *See* Light (natural)

P

Painted art glass, 120–121
Partitions. *See* Glass walls
"Patio" doors, 66, 67–70
Perspective, window placement and, 33
Picture windows. *See* Fixed glass ("pic-
   ture windows")
Pocket doors, 92, 93, 94, 95
Prehung doors, 60

R

Rail-mounted interior doors, 92
Roof windows, 105, 108

S

Sandblasted glass, 74–75
Seasonal affective disorder (SAD), 12
Self-fogging panels, 144
Shapes
   skylights, 117
   windows, 33
Shower enclosures, 155, 164, 165–167
Sidelites
   complete systems with, 58–61
   design ideas, 56–57
   entry doors with, 54, 55, 56–57
Single hinged doors, 52, 70–72
Single-hung windows, 20
Sinks, glass, 154, 161–165
Skylight placement, 106–118

attics, 112–113

bathrooms, 106, 113–114

bedrooms, 106, 107

closets, 116–117

corner rooms, 110

cramped rooms, 114

dining rooms, 108–110

energy considerations, 106–108, 112

entry areas, 110–111

false skylights, 111

fixed skylights, 108–111

hallways, 111, 116, 118, 120

kitchens, 108, 109, 114

large open spaces, 108–110

laundry rooms, 117–118

light tunnels, 116–118

living rooms, 108–110, 112, 113,
  118, 119

mudrooms, 117

practical considerations, 106

roof windows, 108

small rooms, 117–118

staircase area, 111, 120

ventilating skylights, 111–115

Skylights, 100–121

art glass options, 120–121

benefits of, 103–106

blinds/shades for, 115

custom shapes, 117

false, 111

groupings, 102, 103

nighttime and, 103, 106, 107

placement of. See Skylight placement

shafts adding style, 118–119

spreading light, 103

sun control, 115

transforming space, 102, 103

Skylight types, 104–105

custom shapes, 117

fixed, 104, 108–111

light tunnels, 104, 116–118

manual vs. wired openers, 112

roof windows, 105, 108

ventilating, 105

Sliding (gliding) doors, 53, 67, 68, 92

Sliding (gliding) windows, 21

Slumping, 161

Stained glass, 46–47. *See also* Art glass

Staircase

block walls around, 140–141

frosted glass wall abutting, 145

skylight over, 111, 120

Study/home office windows, 38–39

Swinging doors, 53, 67

T

Tempered glass, 85

Tile, glass, 156–159

applications of, 156–158, 163, 173

cutting, versatility, 158

fused glass, 172–173

mosaics, 159

Transom windows

above access doors, 72

above French doors, 58, 59, 80–81,
  88, 89, 96, 97

awning windows as, 21

complete systems with, 58–61

defined, 58

design ideas, 58

highlighting entryways, 54, 55, 70

inside house, 80–81, 86, 88, 89, 96,
  97

mirrored, 81

Tunnels, light, 104, 116–118

V

Ventilating skylights, 105, 111–115

Vitamin D, light and, 12

W

Walls. *See* Glass walls

Window placement/fenestration, 22–23.
See also Light (natural)

  architect determining, 22

  asymmetrical, 22, 36

  bathrooms, 39–41

  bedrooms, 28, 29, 35, 38

  choosing new windows, 30–34

  complementing architecture, 18,
    22–23

  compositions/groupings, 22, 33–34

  dining rooms, 36, 38, 41–43

  environment affecting, 26–29, 33

  existing windows, 30

  factors affecting, 22

  fenestration defined, 22

  high in room, 26, 27, 28, 29, 38

  home office/study/den, 38–39

  kitchens, 35–36

  living rooms, 32, 33, 36–38

  perspective and, 33

  providing scale, 32, 33

  relationship to other openings, 22

  room-by-room, 35–43

  shape options and, 33

Windows, 16–47

  anatomy of, 39

  art glass options, 46–47

  as artistic focal points, 30, 31, 41

  cleaning, 29

  compositions of. *See* Window place-
    ment/fenestration

  connecting outside world, 18

  cross-ventilation with, 19, 85

  as design elements, 18–19

  film coatings, 24, 26

  framing elements, 19, 39, 42, 43–45

  functions, 18–19

  grilles/trim, 39, 40, 41, 43–45

  interior. *See* Windows (interior)

  light color/exposure and, 24–29, 30

  overview, 18–19

  paint/finishes and, 25, 29

  rows of, 22, 26, 27

  seats, 42, 45

  self-fogging panels for, 144

  sills, 39, 45

  structural roles, 19

Windows (interior), 79–89, 94–99

  aesthetic effects, 86–87

  architectural features, 86

  art glass options, 98–99

  benefits of, 79–82, 85

  combining doors with, 94–97

  complementing access doors, 87

  fixed glass vs. operable, 85

  as functional connectors, 85–86

  glass safety, 85

  groupings, 89

  placement of, 85–89

  spreading light within, 79, 84, 85

  structural considerations, 85

  textured glass, 87

  transoms windows, 80–81, 86, 88,
    89, 96, 97

  ventilation with, 85

  window treatments complementing,
    87

Window treatments

  complementing patio doors, 69, 70

  examples, 34, 41

  within glass, 64

  for glass walls, 130–132, 144

  options, 34–35

  self-fogging panels instead of, 144

  for skylights, 115

Window types, 20–21

  arched, 33, 40, 41, 66, 67

  awning, 21

bay, 21

bow, 21, 36, 37, 38

casement, 20, 85

double-hung, 20, 26, 27, 30

fixed glass ("picture windows"), 20,
  25, 26, 27

gliding (sliding), 21

shape options, 33

single-hung, 20